How To Set Up Your
FARM WORKSHOP

How To Set Up Your
FARM WORKSHOP

RICK KUBIK

Voyageur Press

First published in 2007 by Voyageur Press, an imprint of MBI Publishing Company, Galtier Plaza, Suite 200, 380 Jackson Street, St. Paul, MN 55101-3885 USA

MBI Publishing Company titles are also available at discounts in bulk quantity for industrial or sales-promotional use. For details write to Special Sales Manager at MBI Publishing Company, Galtier Plaza, Suite 200, 380 Jackson Street, St. Paul, MN 55101-3885 USA

On the cover: *(main)* It's fun to dream about rebuilding engines and fabricating things from steel, but don't forget that a lot of the work will consist of routine maintenance such as oil changes. *(inset)* Before you invest in any storage solution, do some observation and planning on what works in what situations.

On the back cover: The mechanic's rolling stool increases efficiency by combining a simple centralized tool tray with a comfortable seating platform.

Library of Congress Cataloging-in-Publication Data
Kubik, Rick
How to set up your farm workshop / Rick Kubik.
p. cm.
ISBN-13: 987-0-7603-2549-0 (softbound)
ISBN-10: 0-7603-2549-9 (softbound)
1. Farm shops. 2. Agricultural implements—Maintenance and repair.
I. Title
S675.5.K83 2007
631.3'04—dc22

2006017645

Editor: Amy Glaser
Designer: Brenda C. Canales

Printed in China

CONTENTS

INTRODUCTION

Whether you're starting with a few acres of bare land and a blank sheet of paper, or you're expanding and updating an existing farm operation, planning out the building and equipment for the farm workshop can be a rewarding and exciting task. It's also one that has deep roots in the mechanical age of farming:

> It was the typical dream of these men to have a shop—a personal mechanic's workspace—located in an enclosed structure, such as a garage or barn. A shop enabled the mechanic to come in out of the weather and to continue his work either at night or when it was raining or snowing. . . . These shops were a great source of pride of competency and craftsmanship for these mechanics. These were the men who could figure out what was wrong and fix it. They could repair an engine or a wheel or re-line a set of brakes. They could set the timing, replace the starter, or gap the spark plugs. They could replace the points in the generator or rebuild the carburetor. If the engine ran rough or not at all, they could adjust or fix it. Failure, of necessity, was never an option for them. The machines had to run and they were the ones that made them run.
>
> —from "People and the Land: Elegy, Memory, Promise"
> By Gerald L. Smith and Andrew Gallian,
> The University of the South
> Sewanee, Tennessee

There are good reasons why a key part of every farm is the workshop. Part of it is simply necessity because the farmer needs to be able to fix pretty much anything on the place. Few large farms in North America are close enough to commercial repair facilities to be able to have repairs done there (with the possible exception of tire shops and their specialized equipment). New small farms that have commercial shops nearby might find that these shops are already occupied with work on excavators, heavy trucks, and other industrial equipment, which leaves little room or time for farm jobs. A lot of farmers find they can't or won't go for the prices being charged at commercial shops or dealers, even though those prices are usually quite fair when all is taken into account. The pace of farm work seems to always mean that repairs, maintenance, and modification need to be done at hours when the commercial shops are closed for the day. Those are a lot of good reasons why every farmer prefers to have the kind of shop equipped to repair and maintain any kind of equipment on the place on the farm's own time schedule.

But right along with those business-minded practical reasons, there are reasons that reach back into the long tradition of grit, imagination, and self-reliance that seem to go along with making a person want to farm in the first place. During daylight hours, the farm workshop is the hub of the yard and the first place farm visitors may see on arrival. Whether it's a large commercial farm or the new dream shop of a semi-retired farmer and tractor restorer, it's not hard to sense the "great source of pride of competency and craftsmanship . . ." that goes into the building and equipment of today's farm workshop.

It's with that appreciation of both the practical and intangible elements of a farm workshop that I've put together what I suggest would be useful or desirable to have as equipment for your farm workshop. The reasons for including various equipment and building features reach back right into the late 1950s when I first started to notice the surroundings of shops where I enjoyed hanging around and learning to fix my bicycle. As a child, the first farm workshop I knew was a large, wooden, dirt-floored building built in the days of steam tractors. Rapid technological and size advancements in farm equipment during the 1950s and 1960s made that shop small and obsolete. Nevertheless, the tall ceiling height that had been necessary to accommodate the tall chimneys of steam tractors made it a great place to swing like Jungle Man on a rope hung from a rafter. And the deep layer of fine-textured dust on the floor made for soft landings from accidental falls. Old shops still have their uses, just not the ones you might expect!

The new shop was all breeze block, roll-front doors on separate bays, bright fluorescent lights, and very efficient. As a boy often shanghaied into sweeping the smooth concrete floors, I had plenty of time to observe its equipment and work processes. New technology abounded, such as hydraulic lifts and an electric drill press instead of the old hand-cranked post drill that had caught my brother's finger in its exposed gears—an early introduction to the need for workshop safety.

In 1971, another expansion on the farm meant a new shop needed to be constructed, and it was done with up-to-date planning and equipment of that time. The wooden arch-rib building chosen for that shop was very impressive for the speed of its construction and the amount of equipment it could accommodate in its 40x80-foot space. With minor modifications and upgrades in lighting and insulation, it served its purpose for 35 years until the sale of the place in 2006.

In carrying out research for this book, I've had the chance to visit many farms where I could observe and discuss the elements of a modern shop. Two places in particular stood out as larger and better-equipped than most farm machinery dealerships. I'd like to thank Palin Farms and the Bar None Ranch for giving me a look at some of the impressive facilities that are featured in this book.

Thanks go to the others who have opened their shops to help in this book: Vera and Bill Mokoski of the Treco Ranch, Jerry and Diane Kubik, Pat and Donna Durnin, and Murray Hansen. Thanks also go to the good folks at House of Tools (www.houseoftools.ca), Princess Auto (www.princessauto.com), and Rusteco (www.rusteco.com) for products and photography.

Taken all together, I believe you'll find things in this book to make your farm workshop more efficient and satisfying, whether it's just time for a bit of an upgrade or you're starting from scratch on your dream workshop. As someone still deeply interested in building things and tinkering with machinery, it's helped form a lot of ideas about what's going to be in my dream shop someday. I hope you'll have the same happy result. Good luck and good fixin'!

Rick Kubik, Certified Crop Adviser
American Society of Agronomy

CHAPTER 1

MAKING A PERSONALIZED PLAN

It's tempting to dream of putting up the biggest building you can fit in the yard and filling it with every tool and piece of equipment featured in catalogs, stores, and on television shows, but the result would likely end up being a jumble that's quite hard to use. Plus you'd probably never be able to find anything you need! The best shop for you is one that has the kind of floor plan, equipment, and workflow patterns that maximize the opportunity to accomplish exactly what you want to do. Along with that, the shop would minimize undesired results such as being dangerous, cramped, cluttered, hard to clean up after a job is done, or straining the eyes or back.

All of it comes down to maximizing efficiency, which is shorthand for maximizing your desired results. The first thing you need to define is what it is you desire and what you want to avoid.

Before you start drawing a floor plan, buying tools, or pricing out a building, make a list of what you expect to be doing in the shop. Do you want a shop for routine lubrication, servicing, and repairs, or are you planning to someday handle occasional major overhauls requiring disassembly of engines and transmissions? On the other side of the planning coin, what kinds of annoying/frustrating/dangerous experiences in the workshop can you recall? List those so you can clarify what needs to be avoided.

Your farm workshop will be around for many years, and a well thought-out plan will be ready to handle all your needs in a satisfying way.

NINE WAYS TO NAIL DOWN YOUR NEEDS

1. List any machines that you expect to be working on, in or near the workshop. Don't forget to include ATVs and lawn and garden equipment.

2. List any machines you'd like to acquire within the next 10 years.

3. For each machine, list the approximate length, width, height, and weight, always erring on the larger side.

4. Determine the floor space, door size, and floor strength needed to accommodate the machines on your list. In your plan, it's also very handy to include an outside concrete apron for work in good weather.

5. To start identifying the shape of the shop building, pencil out locations where work can be done without blocking access to the entire shop. For example, if the planter needs extensive work in the shop, where can you place it so it doesn't interfere with running the tractor in for routine maintenance? Lubrication tasks (oil changes and greasing) are a major part of equipment maintenance, so where can you place your lube bay for regular easy access?

6. Identify where you plan to accomplish major repairs and overhauls, if that's part of your expected work. That is where you'll probably want to place specialized tools such as pullers, presses, micrometers, and torque wrenches. It's also a good place for the jacks, hoists, and stands needed to lift and support heavy parts.

7. Once you have a rough idea of where work will most likely be concentrated, make plans to provide plenty of electrical outlets, lights, and ventilation in those areas.

8. For areas on the sketch that look like they might be less used, think about placing storage for those farm tools and equipment you need to have but may only use once in a while, such as digging tools or painting equipment.

9. Drive around and look at other places for ideas on what you might want or need to do in the future.

Next, make a detailed list of what machines you know or expect to be working on at which times of year. This will be another important guide to what you'll need in terms of floor space, doors, floors, lighting, heat, and so on. For example, if you think someday you might want to acquire a combine, it's a good idea to find out how tall and wide they are, and then make sure your shop door is big enough to let a machine of that size pass through. If you're into restoring old crawler tractors, you'd better plan on building a floor slab that can handle the weight of those all-steel beasts.

You can obtain dimensions such as height, width, and weight of tractors and implements from dealers' sales brochures, manufacturers' Web sites, or a quick measurement of a machine at a sales lot. Even if the machine you expect to buy is older or of a different make, the overall dimensions are likely to be fairly similar to those of a similar current machine.

General sizes for farm equipment are also listed in agricultural extension publications such as "Planning Guide to Farm Machinery Storage" by Samuel D. Parsons, P. Mack Strickland, Don D. Jones, and William H. Friday, who are extension agricultural and biological engineers at Purdue University. This publication can be viewed online at www.ces.purdue.edu/extmedia/AE/AE-115.html, or send a note to Purdue University Cooperative Extension Service, West Lafayette, IN 47907, and ask about publication number AE-115.

WELDING AREA SIZE

One of the farm workshop needs that does have fairly well-known size requirements is the welding area. The recommended minimum area for a farm welding shop is 10 feet long and 8 to 12 feet wide. Other things that need to be considered and included are:

Provide space for an oxyacetylene torch, arc welder, welding rods, personal protective equipment (helmets, leather aprons or coats, gloves), and storage for short pieces of metal stock.

Because there will be hot sparks and metal slag flying around, use a fire-resistant wall liner in the welding area.

Keep a large quench tank of water to cool off metal after welding or heating.

In a farm shop, it's convenient to have the welding equipment located near the door because in many cases it will be preferable to extend the cables or hoses so you can do the welding outside for better light and ventilation.

It's fun to dream about rebuilding engines and fabricating things from steel, but don't forget that a lot of the work will consist of routine maintenance such as oil changes.

Being able to easily use the welding equipment outside also prevents moving a large or dirty implement into your shop. Sometimes work-in-progress inside the shop prevents bringing new work into the welding area.

LOCATION ON YOUR PROPERTY

A fully developed individual needs list will help determine the approximate footprint of the shop you want, and then you can think about where it will be located on your farm.

One of the first things to consider is the location relative to your house. The location of the workshop will be a balance between the security and convenience of having a shop near the house, having sufficient room to maneuver tractors and implements, and preserving a handsome, quiet farmstead with good resale value.

Extension publications recommend that a good location is approximately 150 feet behind and to one side of the house. That distance is usually far enough away to protect the living area from the noise, dust, and hazards of large machinery traffic. Engine noise and fumes can be a major irritant if the shop is too close to the house. For example, if you're in your home office trying to do the taxes while a diesel engine is noisily idling just outside, it can be pretty hard to maintain concentration on pages of numbers. Having the shop too close can also mean you have to take more care when maneuvering tractors and implements. Running over a flower bed or backing a cultivator into the side of your

Sometimes you may be the only one on the yard, so a large window facing the house is a good thing to build into your shop plans.

Make sure the location of the shop leaves plenty of room to maneuver trucks, trailers, tractors, and implements in and out of the building.

house is a hazard you can help avoid with proper siting of the workshop relative to the house.

On the other hand, having the shop too far away is not only inconvenient in terms of a long walk to get lunch, but can present a security risk. If you can't keep an eye on the shop from the house and vice versa, it becomes too easy for casual thieves to "grab and go," which is becoming all too common in rural areas.

Another location factor is machinery should not be stored in the same building as livestock because the high moisture levels from animal breath lead to rapidly accelerated corrosion of metal. Hay should not be stored with powered machinery because the increased risk of fire in stored hay means increased risk of losing your machinery in a fire. You may find that your insurance premium increases if hay is going to be stored in or near the shop and machinery.

TAKE YOUR WORK STYLE INTO ACCOUNT

Once the general shape and location are set, it's time to think about how you can arrange the inside layout to make it as efficient as possible. That doesn't have to mean you need something along the lines of a super-neat aircraft overhaul hangar or NASCAR pit where everything is superbly organized and always squeaky clean. If you want to do that, fantastic. However, without the money and helpers those kinds of workshops require, the goal is much harder to achieve on a farm.

On the other hand, there's the ultra-messy "Old Unc's place," a cluttered, greasy lair full of scattered tools and old parts where only the owner can find anything. The owner also probably knows the location and history of everything down to the last bent, rusty cotter pin. Marvels of repair and fabrication can and do emerge from this kind of shop, but it also requires the eccentric genius of an "Old Unc" type to get it really equipped and organized.

If, like most farmers, your work style and circumstances fall somewhere in between these workshop opposites, there are many small things you can do to develop a more efficient shop. These techniques aimed toward continuous improvement are nowadays thought of as the methods of *kaizen*. This concept of continuous improvement is now most associated with the Japanese, but it is one built on the solid foundations of old-school North

Having plenty of room to work inside the shop is important. In this farm shop, there's enough room to bring a service van right up to the machines, plus an access door provided to let a service vehicle in and out while the machines stay in the shop.

Pre-engineered steel buildings supplied as a package are one of the most popular and useful building types today.

American methods of improving industrial efficiency. The relentless introduction of small improvements would have been considered a given way of working in farm shops of days gone by.

If you're of the opinion that you chose farming partly to get away from having to deal with "management-speak" like *kaizen*, consider that you may already be practicing it and not realize it. Take this simple test:

1. The best way to approach a problem is to:
A: Tackle it as a whole.
B: Divide it into small manageable pieces.

2. The best way to build something is to:
A: Make it perfect from the beginning.
B: Make little improvements along the way.

3. When you work, do you:
A: Never stray from the work procedures you've been taught?
B: Think of alternative methods for getting work done?

The more times "B" came up as the answer, the more *kaizen* is something you already practice. Even if you are a normally messy person, there are things you can do to reduce the undesired effects, such as inability to find needed tools and the dreaded periodic cleanups.

Take an extremely typical farm workshop situation: the workbench is messy and disorganized because in the rush of spring planting, you haven't put things away or swept out the dirt that came in on tractor tires and planter wheels. If you simply put everything back in place and sweep up, you aren't practicing *kaizen*. The next time things get busy, you will easily go back and mess up your shop all over again.

If, however, you set up some method that makes it much easier for you to put tools away, put things to be repaired in a place that doesn't interfere with other things, it leads to discarding old, worn-out junk immediately and you're practicing *kaizen*. The key point is to change the method and don't stay with the old way of doing things.

While changing the methods, focus on small changes. Thinking that you need to implement big, costly new strategies tends to inhibit you from fully developing the little improvements that could make your job easier and less frustrating.

A previous generation of ribbed steel buildings offered revolutionary savings over wood construction in its day, but these types of buildings are now used mainly for storage.

For example, in many farm shops, after wrenches have been used on a job, they tend to get tossed onto the workbench rather than put away in the appropriate drawers of the tool cabinet. Nowadays many good quality, reasonably priced socket sets come in a molded plastic case that can be left open on the bench. Even though you and other shop users may still throw sockets and ratchet handles on the workbench, they tend to throw the tools pretty close to where they should be because it's now easy to see and reach the correct place. This small change makes the wrenches easier to find and reduces the time and effort necessary in a more complete cleanup. Another example of

Without an entrance apron, the entrance to the shop may develop problems with the entry of mud and water.

An elevated office area in the shop keeps clear more floor space below for storage or working.

the application of *kaizen* is the type and placement of workbenches, which will be covered more completely in Chapter 7.

All these elements of small improvement should go into arranging a desired interior layout that will guide you toward a final site plan for your workshop. You don't have to necessarily restrict your plan to a simple square or rectangular building. An L- or T-shaped floor plan might work better for your needs, or you might like to have a long narrow building with many front doors. Enjoy using your imagination!

Graph paper and flat cardboard cutouts of equipment can be used to analyze building sizes, shapes, and door sizes/locations for the most efficient storage. If you prefer to do your drawing on a computer, there are many inexpensive architectural or yard software packages that could also be adapted for this task. Be sure to maintain walkways between equipment and working space in the maintenance and repair areas.

Another approach that is perhaps more fun is to go fully three-dimensional by buying farm toys of the exact or approximate dimensions of the equipment you're planning to store and work on. If questions are raised about this approach, you can also reply that it's for serious efficiency planning, not just playing around with farm toys! A very wide selection of these toys are available at most farm equipment dealers. In fact, toys modeled after new equipment have been known to become available even before the actual machines arrive at dealers' lots. You can also research your needs at online outlets such , or the many others that will result from an Internet search for "farm toys."

In terms of moderate cost and wide availability, 1/87 scale is perhaps the most useful for planning purpose. That's because it is compatible in size with the very popular HO scale used in model train building. The scale compatibility lets you use the incredible range of people and animal figures, buildings, and land-forming materials available from model train outlets. One caution to note is when measuring models

There's probably a good scale model available for whatever year or type of farm equipment you want to fit in your shop.

Measure carefully, and don't forget to zero digital calipers before use so that initialization errors don't mess up your calculations.

as a guide to full-size plans, take account of error propagation in calculations. For example, if you're using 1/87 scale models, 1/8 of an inch error in measuring the model is plus or minus 87 eighths, which is nearly a foot at full size. Using a vernier caliper (see Chapter 10) lets you measure models much more precisely to minimize planning errors.

COMFORT AND CONVENIENCE

One of the planning issues for a workshop is whether to include comfort and convenience features such as a toilet, sink, change area, and eating/rest area. If you have employees, these things are required, but even if it's only you that will work there, they are nice to have. For one thing, it saves walking a long way to the toilet and tromping into a clean house with dirty hands, coveralls, and boots.

If you're the kind of owner for whom the workshop is your own ultimate sanctuary and place of creativity, you may also like the idea of having additional comfort features such as a shower, couch, television, refrigerator, computer terminal, sound system, and so on. An office area to talk with visitors, salesmen, or suppliers adds a definite touch of professionalism to the place.

There is, however, an issue that is sometimes overlooked when adding comfort and convenience features: Who is going to clean and maintain them? If you're not really gung-ho about cleaning the bathroom and vacuuming the furniture in your house, are you going to be anymore interested in doing so in your shop? It's something to consider when adding these kinds of extra features; they are nice to have but require upkeep.

Sinks and toilets in workshops tend to get pretty grimy, and most guys don't really care. But if you have business visitors coming by, a pit-like atmosphere may not look too professional.

CHAPTER 2
THE WORKSHOP BUILDING

Whether your dream workshop will be in a new building or a remodeled existing one, be sure to consult local building codes and regulations before you go very far with ordering materials. Since this is a farm building, you may think it's not necessary, but you must make sure before you start. Farm buildings today may be subject to various regulations concerning fuel storage, firefighting access, pesticides handling, and so on. If you forge ahead without any necessary permits, you may have difficulty obtaining insurance or selling the property.

REMODELING

Housing the freshly equipped workshop in a remodeled barn, garage, or shed may save you a lot of time and cost of materials, plus retain the historic look of the farm if that is a consideration. If you're operating a small-scale farm and don't need to move large, modern farm equipment in and out of the shop, this may be a particularly practical and satisfying approach that will not hinder efficiency. The main door openings in older buildings are often big enough to handle the passage of 40 horsepower and under tractors and their implements; under-40 horsepower tractors are the most popular size for small farms.

There are some key upgrades that will need to be done when remodeling outbuildings to make a farm workshop. Perhaps the first upgrade is improving energy efficiency. In days gone by, heating fuel was so cheap that preventing heat loss was a minimal consideration and drafts were

This building is a bit of a fixer-upper, but for certain operations this could make a workshop that works well and stands out from the crowd.

considered something you simply put up with. Air conditioning systems didn't exist on farms during this time.

With today's heating and air conditioning costs, insulation and draft proofing will need to be part of the remodel. Draft-proofing and improved windows will also make the shop much more comfortable to work in. Air conditioning will make the shop a whole lot more pleasant to work in during hot summer months and will keep a lot of annoying insects from getting in through windows and doors left open for cooling the shop. Any heating system, even a wood stove, will likely need to be replaced with a more energy-efficient unit. This will also have the advantage of making the heater less of a fire and carbon monoxide hazard. Stoves that burn corn, other grains, or hay bales allow you to heat your shop with fuels you can grow yourself.

The next concern is the electrical service. The wiring in older buildings may have been sufficient in its day, but it is most likely substandard now. The old tarred-cloth insulation on electrical wires probably will have degraded and may have been gnawed by rodents. When called on to carry

New steel siding can quickly and economically turn a tired plywood-covered building into a modern-looking weatherproof structure.

One economical source for upgraded windows is the local Habitat for Humanity store. Along with good recycled building materials, much of the stock is leftover, never used items from builders and manufacturers.

Dormer windows added to a wooden arch-rib building will bring in plenty of natural light.

For security and working comfort, make the entrance door one of your first workshop building remodeling upgrades.

today's increased loads, old wiring with degraded insulation may become hot enough to start a fire.

Fire resistance in an old building will more than likely need to be improved through the addition of fire blocks in the studding and covering the interior walls with sheetrock. The last thing you need is to lose expensive and perhaps irreplaceable tractors and tools in a shop fire. Fire blocks are simply short pieces of framing lumber nailed horizontally between the vertical studs. When the walls are covered with a flame-retardant material, such as the sheetrock, the fire blocks retard the upward flow of flame and hot gasses between vertical studs and help slow the spread of fire. Check your local building code for recommendations on fire block spacing and sheetrock installation.

Another key consideration is access doors, which are often hard to operate in older buildings that have sagged or settled, or where the doors have rotted or twisted. This consideration applies to both large access doors and ordinary service doors. Working on machinery is hard enough without having to fight with the shop doors every time you need to do something. Doors should be so reliable and easy to operate that you forget about them. They also need to be made secure enough to lock up when you're away. A door that can be kicked in by intruders or torn off by the wind will definitely detract from the efficiency and enjoyment of your shop.

Good lighting needs to be ensured when remodeling an old building. Working on any machinery is much easier and more convenient when you can see what you are doing, and it always helps to have more than adequate light. Most older buildings lack sufficient window area for good natural lighting, and the windows that do exist may be drafty single-glazed units. Light fixtures may be nothing more than a bulb dangling on a wire. An unfinished, age-darkened interior can make the shop into a place much harder to work in due to low light.

Additional windows are relatively easy to install, but they do have the security disadvantage of adding a place for potential intruders to see if there's anything worth stealing, and is another place to break in if they decide there is anything of value inside. Get advice from police, insurance companies, and building supply centers on ways to reduce the security impact of windows. A security benefit of a window facing the entrance to the farm is that it allows you to see anyone coming into the yard.

Also consider flooring and slope for drainage. A strong, crack-free floor (e.g., poured and smoothed concrete) is not absolutely necessary, but it does make the shop much nicer to work in and keep clean. If the floor in an older building was never designed for workshop use, you may be able to get by for a time with something as simple as a good layer of packed gravel, although dropped parts will be harder to find. Essential areas of the workshop can also be temporarily finished with paving bricks or easy-to-lay interlocking rubber tiles. The rubber tiles can also be reused if you subsequently pour a concrete floor, and paving bricks can be reused for landscaping outside the shop.

For concrete floors, a smooth, machine-finished surface is the easiest to sweep up, but the slightly rough texture provided by finishing with a wood float provides more grip when jacks and supports are employed to raise machinery being worked on. Visit other shops if you can to make your own assessment of what is best for your shop and discuss the pros and cons with the contractor who will be pouring your concrete.

For better assurance of dry working conditions, the floor of the building should be several inches to a foot above the outside grade with a suitable entrance ramp leading into the main door. The height above ground level also helps with the task of keeping snow and ice away from access doors. To achieve this floor level with an older building set at grade, buildings in decent condition can be carefully jacked up so that a concrete footing can be poured underneath the walls. To estimate costs, see your telephone directory for "House and Building Movers" or get a referral from a local concrete contractor.

A good selection of doors is available at local Habitat for Humanity stores. Look for a steel-clad garage entrance door or something similar.

Upgrading electrical service in an existing farm building is usually easy because the interior is exposed. This makes it easy to access old wiring and find places to run new wire.

Older heaters may give satisfactory service, but have them checked for fire safety and keep an eye on whether the heating bills may justify a more modern setup.

Hardware and building supply stores sell CO monitors in the same area where fire alarms are stocked. Either battery-powered or shop-current models can be used.

If the outside shell of the building is in fair condition, a coat of paint may be all it needs to protect your remodeled shop from the elements and give it an attractive finish. The new generation of farm building paint is considerably improved in terms of coverage and weather resistance and is often available in custom colors if you want to match a heritage look on the farm. However, the walls and roofs of many farm buildings are now being covered with the new generation of fireproof, weatherproof, colored steel sheets that can be installed quickly and require no maintenance and the time-consuming or expensive labor that goes along with it.

Be careful if you're planning to use steel siding on an older barn that used to have only plain-board siding, especially the type that consists of long vertical boards. With the old wooden siding, some wind likely blew right through to ventilate the barn and help keep the hay dry. Solid siding that eliminates the drafts also increases the wind pressure on the building, which can be a very significant factor on the large wall surfaces of a two-story barn. A building specialist should be consulted to determine if additional bracing is required with solid siding.

NEW WORKSHOP BUILDINGS

Planning a new building for your new workshop opens up many new options in layout and type of construction. Pole

Rubber tiles provide a comfortable working surface, whether it's over concrete, gravel, or packed dirt.

frame, metal frame, wooden/metal arch, masonry block, or conventional construction have all been used for farm workshop buildings. More recently, heavy-duty fabric-skinned farm buildings have become common. Straw bale construction uses a material you may already have in abundance on your farm. One shop I visited that had a particularly comfortable working atmosphere was at a potato farm that had the workshop partially set into the ground. The mass of soil against the walls kept it comfortably cellar-cool all year long.

There is no general rule as to which type is most economical or suitable for any one situation. It is often not so much the building itself but the upgrades such as windows, doors, and floor types that cause the biggest variations in price.

The simplest and cheapest type of fixed workshop building is usually the basic pole shed that is open on one side. The openings between the front support poles form natural bays for equipment. To keep initial costs under control, you can enclose only one end of the shed and add doors over time to enclose more bays.

A key point in any type of building is the design load needed to withstand factors relevant to your area. These include side load from wind, roof load from snow, ability to withstand extreme storms, and more. Do the research yourself or work with a knowledgeable local engineer, architect, or contractor.

If the shop building is to form one side of a livestock enclosure, this also creates issues that should not be ignored either. Hogs will root deeply under the footings in short order. Cattle will rub against the walls and may create holes. Horses may chew wooden walls. For these and other reasons it may be wise to put additional reinforcing on walls that will be exposed to livestock.

More of the farm buildings constructed today are pre-engineered structures available from manufacturers or contractors. These package buildings take advantage of the benefits of standardization and allow a rapid erection of a building with the features you've determined you need in your new farm workshop.

If you plan to put up the building yourself, be realistic about your abilities and the timeframe for construction. Many projects rapidly fall behind schedule due to reasons such as spells of bad weather or not budgeting the time needed for other commitments. Ask your neighbors and

A long, narrow multi-bay workshop lets you store tractors or other equipment without interfering with work going on in another area.

Modern straight-wall buildings use strong vertical members spaced wide apart with less expensive horizontal stringers to provide attachment points for siding.

experts at building supply centers for as many examples as you can get of how long it actually took to complete buildings. By having a large number of actual local examples, you can form a more realistic idea of how long it will take to get your building finished.

When having the building erected by a contractor, it is up to you to be diligent about selecting one you feel is the best for your situation. Do not rely solely on a low-ball price, family connections, or advertising, alone. Check with friends and neighbors who have dealt with construction recently and at professional building supply centers, and take their comments about contractors into account. If a contractor can provide a list of previous customers as references, check with the references about workmanship, timeliness, and adherence to contract terms.

Once you've selected a contractor, the fun has just begun. While most dealings between owners and contractors result in satisfactory completion of the desired building, there are always difficulties that occur, due perhaps to the differing views of your project.

To you, the owner, this is the most important project in the world and needs all the attention it can get at every opportunity. You are eagerly looking forward to when the building can be put to its intended use. To the contractor,

End walls in arched building design can be modified with windows, fans, and doors to suit particular workshop needs.

this is more than likely one of the many projects on the go. The contractor is looking forward to best management of time, labor, and equipment over the construction season to achieve overall business survival and growth. The differing views may mean your project could sit idle for a few days during good construction weather. From another perspective, satisfyingly rapid activity at your site may leave another owner wondering why his/her site sits idle during good weather.

When frustration and harsh words result, it can generally be traced to poor communication between owner and contractor. That's why it is so important to select a contractor you can trust, even if it means paying a little more. If you're not certain from the start that you trust your contractor, it's surely going to be difficult or impossible to communicate clearly, firmly, and on a regular basis.

A clearly written contract is also an essential first step in having both you and the contractor clear about what is expected on the job. It may not be entirely sufficient, though. Even if the contract is eventually completed, disputes that delay completion mean a construction season has slipped away and you may be left with bitter feelings that reduce your enjoyment of the new workshop. Nevertheless, a clear contract with a trustworthy contractor goes a very long way toward ensuring your new building goes up as hoped. In the contract, some items to specify in writing include:

Duties of the contractor. The contractor supplies all labor, equipment, and materials to complete the structure. Spell out the time schedule for completion.

Duties of the owner. Specify if any of the work, equipment, or materials is to be supplied by you, such as rough site grading or provision of utilities during construction. If you as the owner take on any of the work, including grading the site to save money, be extremely diligent about measuring the correct site elevations, grades, and distances from property lines or other land features. A seemingly small error here can lead to years of headaches. Hiring a professional land surveyor should definitely be considered.

Drawings and specifications. Farm workshops sometimes were planned out on a brown paper bag or built without plans at all, but those days are long gone. No building should be constructed without a complete set of drawings and written specifications included as part of the written contract.

Changes and substitutions. It's rare to have a building that doesn't undergo some changes from original plans and specifications. You and the contractor need to agree on

Traditional track doors can provide long service if the tracks and rollers are protected from dripping water by a roof overhang. If water does get into the track, inevitable corrosion and wood rot makes it harder to use.

procedures to be followed and who is to pay for any differences in cost.

Laws, permits, and regulations. Construction must conform to all applicable laws and regulations, not only for the safety and enjoyment of the building, but to make sure it is insurable and will not cause difficulties when reselling the property. Make sure the contract specifies who is responsible for obtaining any necessary permits, inspections, and final approvals.

Insurance during construction. Make sure the contract spells out what is necessary for adequate risk protection during construction so that you do not face liability in case of injury to contractor's employees, property damage from contractor's activities, and so on.

Site access. Many new farm buildings are sited where large trucks do not have easy access to deliver heavy loads of materials such as concrete, structural steel, or roof trusses. Lack of a good road may become especially important if rains make the site soft and muddy. Make sure the contract specifies who is responsible for maintaining site access and who bears the extra cost if materials have to be unloaded off-site and brought to the site in smaller vehicles.

Storage of materials. Secure, weatherproof (if needed) on-site storage of construction materials may be either a contractor or an owner responsibility. Specify this in the contract, as well as what to do if there is a dispute over what type of storage is needed.

Cleanup. Upon completion of the construction, someone must clear the site of all construction debris and clean up building surfaces. Specify whether this is the responsibility of the owner or the contractor.

Payment schedule. Farm workshop building projects will generally be of a size that an initial deposit and final payment on completion are all that will be required. But if it's a large project with complex interior fittings or expensive equipment, interim payment of portions may be needed. Make sure the contract addresses this issue.

Warranties. Terms of the contractor-supplied warranty should be spelled out in the contract. Provisions also should be made for transferring to the owner any warranties provided by manufacturers or suppliers of component parts.

Service manuals and operational instructions. The contractor should be responsible for providing the owner with operational and service manuals for component equipment, such as heaters.

WINDOWS IN NEW WORKSHOPS

Good lighting is essential in a workshop, and natural light through windows provides a source free of the many shadows involved with overhead or point-source lighting. When windows are used for illumination, keep the window area equal to at least 8 to 10 percent of the floor area.

For all their advantages, windows also have the security disadvantage of adding a place for potential intruders to see if there's anything worth stealing, as well as a place to break in. Get advice from police, insurance companies, and building supply centers on ways to increase the security impact of windows. A security benefit of a window facing the entrance to the farm is that it allows you to see anyone coming into the yard. For other windows and from a security standpoint, skylights may be a good alternative.

DOORS IN NEW WORKSHOPS

In new farm buildings, the door choices include horizontally sliding doors, accordion-type doors that fold to the sides or top, and a heavy-duty overhead door (similar to typical home garage doors).

Wind can be a big problem for horizontally sliding doors. While the top is usually firmly secured in a track-and-roller arrangement, the bottom is only held by short vertical tabs, usually not the full width of the door. In a strong wind, a fully opened door may burst out of the lower tabs and flap itself to destruction. Adding bigger, stronger locating tabs at the bottom of the door is usually not a viable option because the lower track can fill up with dirt or ice. To prevent wind damage, keep this type of door closed when not in use or install latches at chest height to hold the door firmly in place when it's left fully open.

Workshop entrance doors are the usual type of steel-clad door found in garages, with the addition of slightly wider width (36 inches or more) to accommodate large objects without opening the main doors. Install a quality deadbolt lock to keep your workshop secure.

INSULATION

Whether the building is remodeled or new, insulating it is key to achieving a comfortable working environment while

When the door is unlatched and slid open, guides are needed to secure the bottom against unintended movement that will damage the rollers or the door.

Entrance doors can be built into larger doors to improve accessibility.

keeping heating/cooling costs down. Insulation specifications are based on R value. The R (resistance) rating is a standard building industry rating of how well the insulation system resists the flow of heat from the hot side to the cool side. The higher the R value, the more effective it is (e.g., R15 insulation retains heat better than R7). Panels or packages of insulation material will usually have the R rating printed right on the package. A structural arrangement also has an R rating. For example, a properly built 2 x 6 wall insulated with fiberglass batts has a "whole wall" rating of R14, but that can vary depending on how well the wall is constructed.

Since insulating materials are relatively lightweight and the installation skills required are not too hard to learn, insulating the farm workshop is a job well within the do-it-yourselfer's realm as long as you wear appropriate protective equipment. Glass fibers are very itchy on the skin and they are not the kind of thing you want to breathe into your lungs, either.

Check with building supply centers for up-to-date recommendations on what materials and R values are best for your area. Local codes will list recommended R values for buildings, but one thing to remember is that local building codes almost always list lower R value requirements than the Department of Energy (DOE) recommendations.

HEATING, VENTILATION, AND AIR CONDITIONING (HVAC)

To maintain a comfortable working environment in many areas of the country, your farm workshop will require suitable heating and cooling. Fans and vents are also needed to keep fresh air circulating and draw off fumes and smells.

A hangar-type bi-fold door rises completely and is held securely on side tracks throughout its movement.

An inside view of a vertical bi-fold door shows the heavy steel side frames and electric door lift.

As a start, a lot of the ventilation needs may be taken care of by leaving the main access doors open. But this unfortunately also lets in flies, mosquitoes, rodents, birds, and other bothersome pests. Fans and vents should be used to maintain good working conditions within the shop.

When cooler weather comes, a heating system will also be needed. A heated floor system where warm liquid is circulated through tubes embedded in the concrete floor is very nice to have if the shop is continuously used. This type of heating does take a while to warm up the building. A normal space heater or forced-air heater is also desirable for quickly warming the shop. Modern stove designs that use wood, corn, small grains, or straw bales as fuel offer a way to heat the shop with materials harvested from the farm. Whatever kind of heater you choose, find out the proper installation and operating procedures to minimize hazards from carbon monoxide.

An inside view of a vertical bi-fold door shows the heavy steel side frames and electric door lift.

ELECTRICAL WIRING

A modern farm workshop needs electrical service capable of safely operating large power tools, an air compressor, and

Heated forced air remains a preferred way to quickly warm workshops and, in conjunction with ceiling fans, to maintain a comfortable working temperature at floor level.

Gas-fired radiant heating tubes are an increasingly popular way of heating workshops.

welding equipment. The minimum recommended is a 200-amp, 240-volt service. Check electrical codes or consult electrical contractors to find out what's recommended in your area. It's a lot easier and cheaper to have it installed during construction than it is to upgrade at a later date.

Unless you are thoroughly skilled at installing proper wiring and have the right tools, always have a qualified licensed electrician do the work. Even a 120-volt house current can cause very dangerous electrical shock, and improper wiring often is the cause of serious fires. Wiring should be approved by an electrical inspector before the service is connected. If this step is avoided there may be later complications with insurance coverage and payouts.

Install 120-volt duplex electrical outlets at least every 4 feet along workbenches and at least every 10 feet along the walls. If power tools are intended to be used away from walls, it may be best to install outlets in the floor or suspend them from the ceiling so that you don't need to run cords across the floor.

A 240-volt, 50-amp or larger outlet is needed for the welder. Have at least one welder outlet near the main entrance door so welding can be done both inside and outside. An additional outlet outside the shop can be useful when working on machines inside the shop.

Ground fault interrupters (GFIs) are recommended on all single-phase 15- and 20-amp convenience outlet circuits. The National Electrical Code requires GFIs for 15- and 20-amp receptacle outlets installed on the outside of a building or near wet areas.

LIGHTING

Well-selected lighting in your workshop makes it much more efficient and enjoyable and prevents problems with glare, flicker, and more. Consult local codes and ask building suppliers about recommendations on built-in lighting. In general, fluorescent lighting is only recommended for heated workshops because in cold shops, fluorescent lights are extremely slow to start. Mercury vapor or high-pressure sodium lamps are also slow to come on, so if you're planning to use them for general lighting, plan to install a few incandescent lights for quick response.

No matter how well lit the workshop, the odd shapes you'll be encountering in the farm workshop mean that point-source task lighting will also be needed. The old standby mechanic's trouble light is always needed to bring illumination into and under machines. Recently these lights have become available in fluorescent, which are more energy efficient, and throw off much less heat, making them more comfortable to work with.

FLOORS

A 4- to 6-inch-thick concrete floor built on a well-drained, compacted fill base should be sufficient for most farm workshops. To help resist cracking, make the floors thicker a distance of 2 feet or more in front of doors where heavy machines will be entering. Concrete floors are usually laid in sections separated by flexible joints so that any shifting in the underlying soil does not lead to cracks in the concrete.

Cracks in the concrete aren't just an engineering issue—ants love to invade shops with cracked floors. When you are lying on the floor doing repairs it is very unpleasant to have an assault force of ants swarming into your coveralls.

BUILD IN FIRE SAFETY

When constructing a farm workshop, the site plan should allow for adequate spacing from any other buildings to prevent the spread of a possible fire.

Locate the shop at least 40 feet away from aboveground fuel storage tanks to minimize the potential for spread of fire. Consider greater separation if the buildings are in line with prevailing winds.

When constructing a new workshop, build with fire prevention in mind by including features such as fire doors, a firewall between hay/bedding storage and stabling or work areas, flame-retardant or fire-resistant materials, fire-retardant latex paint, smoke detectors, fire alarms, and automatic heat-sensitive sprinkler systems.

Battery-powered emergency lighting allows the evacuation of people and animals in case of power failure, and a water source on or near the premises is very beneficial if there is a fire. It gives the fire trucks a very close water source and saves precious time from waiting for the water tanker to arrive on the scene.

A floor drain built in at the start makes it possible to wash down equipment in the shop—away from wind and cold.

This shop-built solar panel provides all the heat needed for a 10,000-square-foot commercial shop except when temperatures occasionally dip below minus 30 degrees Fahrenheit. Thankfully, that's not many days of the year.

CHAPTER 3
COMPRESSED AIR SYSTEM

One of the key elements in your new shop will be a compressed air system to run a variety of tools such as wrenches, grinders, sanders, buffers, and drills. The same system will also provide compressed air for filling tires.

Air tools have several advantages in a modern farm workshop. They are generally less money to buy than electric-powered tools. Air tools are also lighter in the hand than electric tools because there is not the weight of an electric motor's armature and windings in the air tool; it is just a small turbine wheel.

The first step in installing the system is choosing a suitable air compressor and air tank. Generally the recommendation is to decide based on the air requirement of the largest air tool. In the specifications for any air tool, you'll see listed a specific average air consumption such as 4.5 cfm at 90 psi. That translates to in average working conditions this tool needs a supply of 4.5 cubic feet per minute of air compressed to 90 pounds per square inch. Average cfm is typically based on a 25 percent duty cycle (e.g., 15 seconds operation per minute total time). This is fine for tools such as nailers and impact wrenches that are used more or less intermittently. But for tools such as grinders or sanders that are used on a more continuous basis, it is a good idea to multiply the average cfm times 4 to get a continuous cfm rating.

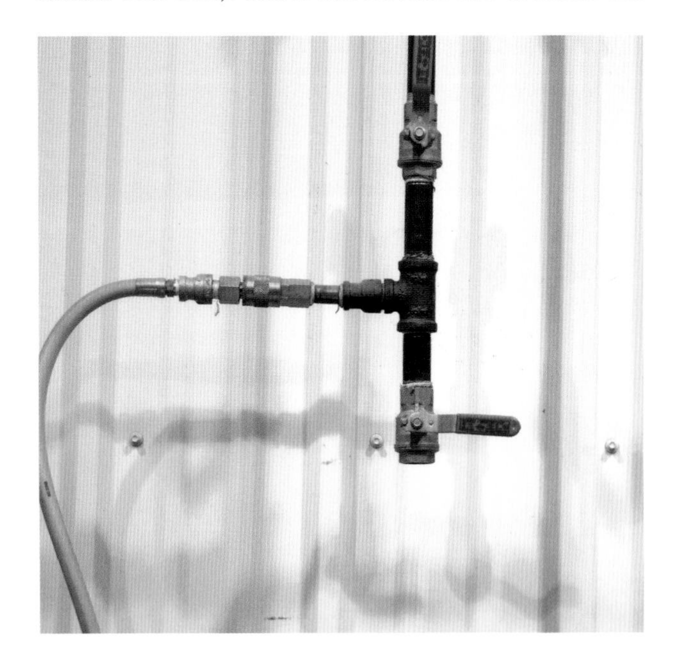

A compressed air system is a very useful and powerful workshop tool and can remain so for decades with a little attention to handling moisture in the system.

Typical air tool consumption, cfm at 90 psi	
Angle Disc Grinder, 7-inch	5–8
Chisel/Hammer	3–11
Cut-Off Tool	4–10
Drill	3–6
Dual Sander	11–13
Grease Gun	4
Impact Wrench, 3/8-inch	2.5–3.5
Impact Wrench, 1/2-inch	4–5
Impact Wrench, 1-inch	10
Mini Die Grinder	4–6
Nailer, framing	2.2
Nailer, brad	0.3
Needle Scaler	8–16
Nibbler	4
Orbital Sander	6–9
Ratchet, 1/4-inch	2.5–3.5
Ratchet, 3/8-inch	4.5–5
Rotational Sander	8–12.5
Shears	8–16
Speed Saw	5

Other Compressed Air Equipment Consumption	
Sandblaster, cabinet	5 -15 cfm at 80–120 psi
Sandblaster, benchtop	5 -15 cfm at 120 psi
Sandblaster, pressure	10–80 cfm at 100 psi
Sandblaster, siphon	4.5 cfm at 90 psi
Sandblaster, touchup	6 cfm at 80–120 psi
Paint gun, siphon type	8.5–12 cfm at 30–70 psi
Paint gun, gravity feed	12–15 cfm at 30–60 psi
Paint gun, HVLP type	14–17 cfm at 30 psi

In a farm workshop, it is possible to have an air system that supplies no more than the capacity of your highest demand air tool. After all, no matter how much your range of air tools multiplies in the future, in a farm workshop you will likely be using only

Locating the compressor or an air line near the main door provides compressed air for use in or outdoors.

Easy-to-wind-up hose reels keep your workshop from having an unsightly and unsafe tangle of air lines snaking all over the floor.

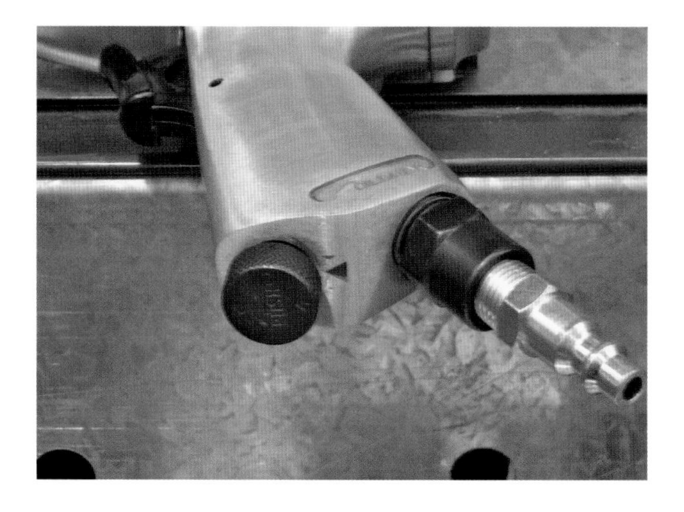

Many air tools have a torque regulation control near the inlet. It's not a precise setting, but it does allow you to turn down the force so you don't accidentally twist off the smaller fasteners.

The pistol-type impact wrench is the most familiar air tool because of its widespread use for quickly loosening and tightening wheel lug nuts. It makes short work of fasteners that are too tight to remove with muscle-powered tools.

one tool at a time. That's different than an industrial shop where a number of workers may be using air tools at once and the size of the compressor and tank needs to be scaled up to match.

However, your working environment is also a good reason to go with a larger air compressor than is needed to run your largest air tool. If a smallish compressor is used, it must run almost continuously to supply the tool with enough air. You can observe examples of this at any construction site where small compressors are used because they are more portable. The compressors need to run almost continuously to run nailers and other tools. While that's acceptable outdoors, in an enclosed shop it can make for a shop environment that is unacceptably noisy.

A higher-capacity compressor pump, such as a dual-stage twin-cylinder, and a larger air tank are usually preferable inside a shop because it needs to run less often. Multistage units tend to be more efficient and compress air to a higher pressure while generating less heat and extending compressor life. A large upright tank minimizes the floor space needed. For maximum noise control, the compressor may even be located outside the main shop, perhaps in an adjoining storage building. Air lines of the right size and material can be installed to deliver air where needed. Key issues when deciding which air compressor to choose also include life expectancy, frequency of use, relative noise level, and warranty. As a rule of thumb, better compressors will also have a longer warranty and up-rated frequency-of-use rating such as "industrial" or "extreme."

With the recent wave of farm consolidations into larger units and movement of many small industries to overseas

Because the impact exerts such concentrated twisting pulses, it can damage regular sockets. Heavier impact sockets are needed. Note the cutaway corners to prevent damage to corners of nuts and bolts.

A big impact wrench for the big jobs is used with a D-grip handle to help keep control under the high torque exerted.

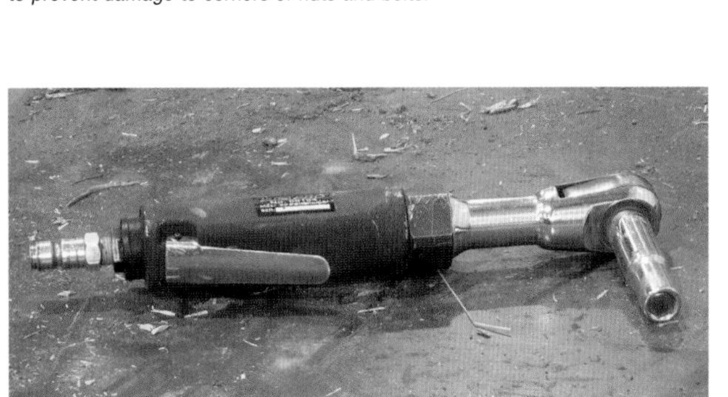

An air ratchet wrench exerts only twisting force, like a normal hand wrench, instead of the twisting and hammering of the impact wrench. Repetitive work can be sped up quite a lot with the air ratchet.

The air die grinder is a much more robust version of the electric rotary tools used for hobby work. The small size of this tool makes it easy to get in to grind in tight places, such as where a fastener has broken off deep in a machine.

production, large air compressors are now often available at farm and industrial equipment disposal sales. If you're in the market for a new or larger air compressor, it's worth knowing system prices so you can look for deals at farm and industrial equipment auctions.

MOISTURE

All air taken in for compression contains some level of moisture, even in Arizona. When the air is compressed, some or all of the moisture condenses into liquid form, which can cause rust and scale inside the tank. Liquid water in the compressed air can also make its way into tools and cause corrosion and interfere with lubrication and cause premature failure.

Moisture traps and drains are provided on the air tank itself and should be installed in the piping that carries air to other points in the shop. Open the drains daily or install automatic drains. Air dryers can also be installed to further reduce condensation issues in tanks, lines, and air tools.

AIR LINES

You can, for a start, get by with a flexible air hose to the tools, but having lines lying all over the place makes your shop look cluttered and creates a tripping hazard. It also increases the problems of hoses getting tangled up in equipment, making it difficult to pull the hose around to where you need it. Air-hose storage reels, self-retracting coiled hoses, or installation of air drop lines above the work area help eliminate these problems.

For an air distribution system within the shop, most farm workshops use threaded carbon steel (commonly known as black iron) plumbing pipe. This type of pipe has the advantage of being reliable, readily available, relatively

The air die grinder is a much more robust version of the electric rotary tools used for hobby work. The small size of this tool makes it easy to put a lot of grinding power into tight places, such as where a fastener has broken off deep in a machine.

The high speed and compact size of the air cutoff saw can cut through thin sheet, as well as thicker pieces that a shear could not handle.

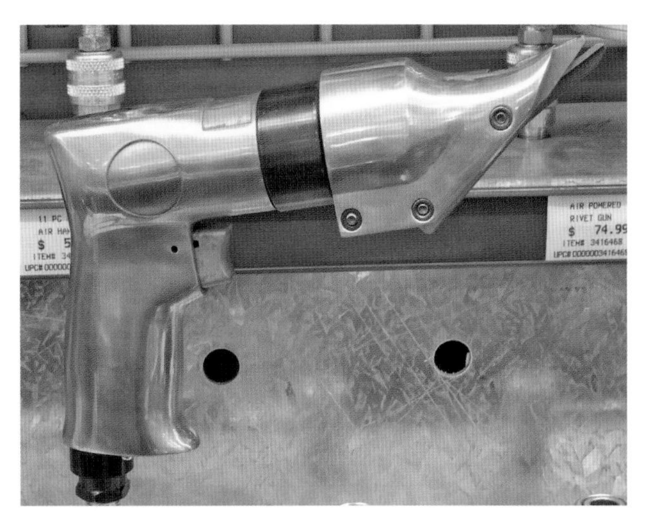

An air-powered version of scissors slices through sheet metal in straight or curved lines.

The air chisel (not pictured) can be equipped with a crimping attachment to form L-shaped creases along the edge of sheet metal panels in preparation for joining by welding. The crease allows one panel to overlap the other even though the top surfaces are flush with each other.

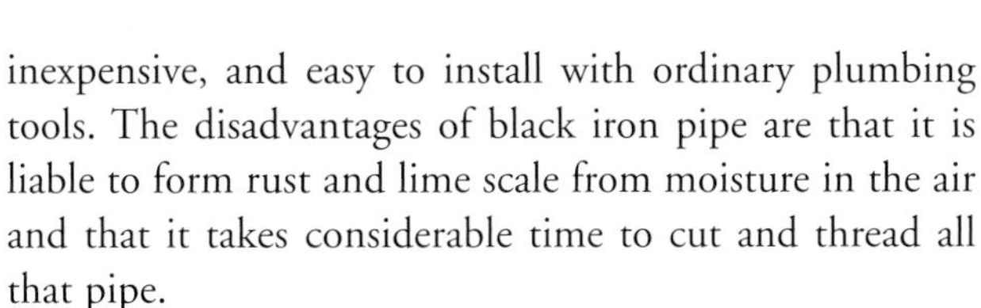

inexpensive, and easy to install with ordinary plumbing tools. The disadvantages of black iron pipe are that it is liable to form rust and lime scale from moisture in the air and that it takes considerable time to cut and thread all that pipe.

Galvanized or copper plumbing pipe are sometimes used instead of black iron pipe. Either one is much more resistant to corrosion, but it is also much more expensive if bought new. Compared to threading iron pipe, soldering the joints on copper pipe is also a job that takes more skill.

With plastic pipe, only certain kinds can be used for handling compressed air due to risk of the pipe exploding. If you are considering plastic piping, use only a material specifically rated for gases under pressure. Check with your supplier and investigate local building codes before laying out any money for plastic, compressed air line material. Rigid PVC or ABS plumbing pipe is not a good choice and may actually be illegal according to local building or insurance codes. The reason is that if a rigid plastic pipe cracks or suffers joint failure while holding air pressure, the result can be an explosive scattering of pieces around your shop.

Polyethylene (PE) pipe has high resistance to fracture propagation (crack spreading), which limits the extent of fracture in case of accidental damage. The light weight of PE pipe makes installation relatively quick and easy. Be especially sure the joints are properly glued or threaded and consider expert installation if you do not have experience with the material. Joints are often the weakest link in any plastic piping system.

Bits for the air chisel allow it to perform a variety of chipping and cutting tasks that would be very slow if undertaken with a chisel and hammer.

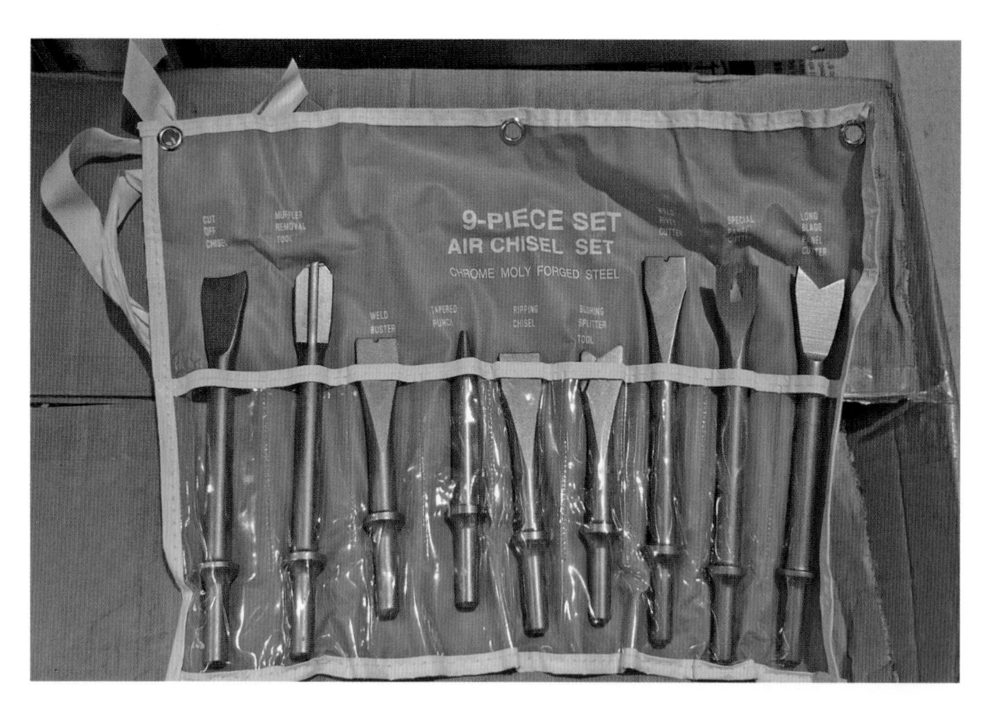

Even though a large compressor is needed for the compressed air system, it's also nice to have a small, easily portable compressor for jobs where it's easier to string an electric cord than to run out a long air line, such as at the tops of grain bins.

Another useful approach to portability in compressed air systems is to have a generator and compressor on their own cart, allowing you to take air power anywhere.

There are also pre-manufactured modular kits for compressed air distribution. These kits use powder-coated aluminum tubes joined together with special push-in fittings. This provides secure and airtight connection without any special tools, soldering, gluing, or threading pipe, and it looks a lot nicer. The layout can be expanded or rerouted if necessary because all the fittings are removable and interchangeable. You could also disassemble the system and move it to a new location if you expand the shop or move.

Whatever system you use, pay close attention to achieving airtight joints. Air leaking out through piping joints is a common loss of efficiency in compressed air systems.

Air Tools and Uses

Air tools are available to accomplish most of the workshop tasks associated with hand-powered or electric-powered tools. Air tools are fast and easy to use, but don't forget that they make it just as fast and easy to damage or destroy the work if used too aggressively.

CHAPTER 4
WELDING AREA

Even a small farm generally has a regular menu of metal repairs that need to be done on everything from tractors to metal gates. A welding rig can often pay for itself in short order by either eliminating the cost and time needed to get a new part or keeping an older machine running when new parts aren't available.

Every farm workshop is different so to make the decision on welder size and type, get advice from a skilled welder or at reputable welding supply shops, and take welding courses offered in your area at colleges or welding supply houses. Another useful reference is the book *The Farm Welding Handbook* by Richard Finch. It contains very good information on all aspects of farm welding choices and operations.

One item that does bear repeating is the safety warning that welding fumes can contribute to the development of Parkinson's disease, which affects the brain and nervous system and causes symptoms including uncontrollable tremors (shaking), slowness of movement, stiffness, difficulty with balance, and depression. This problem is more likely to affect professional welders who are exposed to fumes all day, but why take a chance with such a debilitating consequence? Make sure your welding area is well ventilated.

TYPES OF WELDING EQUIPMENT AND USES

In the not-too-distant past, most farm workshops were equipped with two types of welding sets: oxyacetylene (gas welder) and an arc (AC electric stick) welder because they were what was available at the time. That's why those are the types of welders you'll most often see at farm auction sales. Both types still have qualities that keep them useful on a more limited basis in today's farm workshop.

A strong steel bench is needed for electrical conductivity, heat resistance, and the ability to withstand hammering, which are all regular parts of welding.

TEMPERING AND ANNEALING

The metal-heating role of the oxyacetylene torch is particularly useful in the farm shop. At its simplest, heating metal makes it easier to bend it back into its original shape or into a new shape. Pieces may also be heated to loosen parts or swell the metal to allow the insertion of a friction-fit piece.

However, whenever you heat and cool metal there will be changes to its ductility (ability to be shaped), strength, and resistance to wear. This is why knife blades are heated up for initial shaping then specially heat-treated to enable them to hold a sharp edge.

For example, with steel, heating it and then letting it cool down in air will soften and weaken (anneal) it to make it easier to shape by bending or hammering. Heating and then cooling rapidly (quenching) can restore much of the original strength and resistance to wear.

Annealing and tempering is useful when you're working on a spring or cutter blade. A rough guide for the steel you'll typically encounter in farm equipment is that heating to a bright cherry red and letting the part cool off naturally makes it soft enough to shape and move. Reheating again to the same bright cherry red and quenching in water hardens the steel again.

With non-ferrous metals such as a copper sealing ring or an aluminum irrigation clamp, the process is reversed. Heating to a temperature just below melting and then quenching makes the metal softer and easier to shape. Reheating to a temperature just below melting and then letting the piece cool naturally in air re-tempers the metal. Various aluminum alloys may also require a longer heat soak period, such as heating to about 500 degrees Fahrenheit for an hour or so, then natural air cooling.

The above is a quick overview of the annealing tempering process as an example of the place for the oxyacetylene rig in your farm workshop. Reference books on gunsmithing, knifemaking, or art metalworking are good sources with more complete details.

A welding bench with a commonly needed saw and vise are in close proximity.

Never leave gas welding sets as shown above; always store and transport gas welding sets in an upright position.

Today, the consumer availability of gas metal arc welders (MIG and TIG welders) and plasma cutters have added new choices and new capabilities for a farm welding shop. Which type should you buy for your farm workshop, oxyacetylene, stick, MIG, or TIG welder? The short answer is, plan on eventually getting them all because the highly varied nature of farm welding brings up a number of jobs that each welder handles best.

- An oxyacetylene set can commonly be found used and at a farm auction, and new sets run about $250 plus cost of gas refills (about $30).
- Heavy-duty stick-type arc welders are also commonly available at farm auctions.
- Allocate most of your resources to acquiring a new, modern wire-feed MIG or TIG welder. They can often be tried out in the demo rooms at welding supply shops or farm equipment shows.

Regulators on the tops of gas welding/welding set bottles are precision flow regulation devices, so don't crank the handles down hard when shutting off gas.

A flow of acetylene is lit off first, which produces the characteristic feathery yellow flame with a lot of soot.

• If your operation includes a lot of metal cutting for work or hobby reasons, put the plasma cutter on the "equipment to get" list.

The following is a brief overview for each setup in your workshop.

GAS WELDER (OXYACETYLENE)

The oxyacetylene torch mixes oxygen and acetylene gases to produce a flame hot enough to melt steel. The flame can be adjusted for different tasks by changing the ratio of the volume of oxygen to acetylene. By using a special torch head, steel can be cut by preheating the metal and then introducing excess oxygen to the mix.

In the farm workshop, the gas welder is actually used very little for welding steel, although it is quite capable of doing so in a pinch. Electric arc machines (stick, MIG, or TIG) can weld steel at a lower cost and with less skill required.

The chief remaining roles for the oxyacetylene welder are heating metal to prepare for welding or to loosen stuck pieces, cutting with the torch attachment, and brazing of thin metals, cast iron, and dissimilar metals (e.g., brass to steel).

The gas welder is portable for those times when you need to cut, heat, or weld in the field or other places far away from the workshop. Just remember to always transport the gas welder in an upright position. If the acetylene bottle is laid horizontally, acetone can collect in the regulator and cause an extremely dangerous fire when the torch is lit.

ARC WELDER

For a relatively low initial cost, the arc welder provides an inexpensive method for welding repairs on farm machinery and structures and applying layers of tough steel (hardfacing) to soil openers and loader bucket edges. The name "stick welder" came about in reference to the sparkler-like electrodes (sticks) that are gripped in the electrode holder to produce the arc and melt into the weld.

Electrodes are gradually used up during welding so the welding process must stop while a new electrode is installed. To get maximum use from electrodes,

As oxygen is slowly introduced into the gas flow, the flame begins to burn much hotter and more completely, producing a noisier blue flame.

experienced welders stop the weld with the stub end of the used-up electrode protruding from the weld as it cools. When a new electrode is installed, the arc is struck at the end of the old stub and allowed to cool so the stub end is fused to the new electrode. The whole thing is then gently wiggled loose from the weld and the arc is re-struck to continue welding.

If considering a stick welder, look for a heavy-duty AC/DC model. For most applications, DC reverse-polarity welding provides easier starts, less sticking, less spatter, and a better ability to weld thinner material.

Since a heavy-duty welder needs 220-volt wiring, the work usually must come to the shop unless you have a large portable generator. An advantage for the stick welder is that it permits work on steel that is somewhat rusty or dirty, or is situated in somewhat windy outdoor conditions often encountered in farm welding jobs.

Using a stick welder for thin materials or any metal other than steel can be done, but it takes a very skilled operator, the right electrodes, and plenty of time. MIG or TIG welding sets are most often a better choice.

MIG WELDER

The MIG welder is also based on the heat produced by an electrical arc, but it is different from the stick welder in two key ways. One is that the arc is shielded by an externally supplied gas that does not react with the metal being welded. Hence the name, metal inert gas (MIG) welding. New developments have led to availability of choices in both inert and reactive gases to shield the arc so the "inert" name is not exactly correct anymore. But the welding equipment is still often called a MIG welder in ads and stores.

The second difference is that the filler metal is continually fed to the tip of the torch so the operator can keep the torch tip a constant distance from the work and make long welds with fewer starts and stops. With stick welding, the operator has to learn the skill of

Copper sealing rings are often used as gaskets of relatively soft metal between steel parts. With time and vibration, these rings can harden and lose their ability to seal. They can be softened with heating and cooling and treated with a copper-depositing spray (right) or a copper-rich high-heat silicone (left).

constantly moving the holder closer to the work as the electrode is consumed and know when to stop and manually change electrodes.

In the farm workshop, the advantages of a MIG welder are:

- Easiest welding process to learn. Usually even a first-time MIG user can achieve a good-quality weld.
- Ability to weld a wide range of thickness of metal, from light-gauge sheet to thick plates.
- Can weld all common metals (carbon steel, stainless steel, and aluminum) in thin material or thick plates.
- Speed of operation up to four times faster than stick welding.
- More efficient use of filler: 50 pounds of MIG welding wire yields approximately 49 pounds of metal deposition; 50 pounds of stick electrode rods yield approximately 30 pounds of deposition.

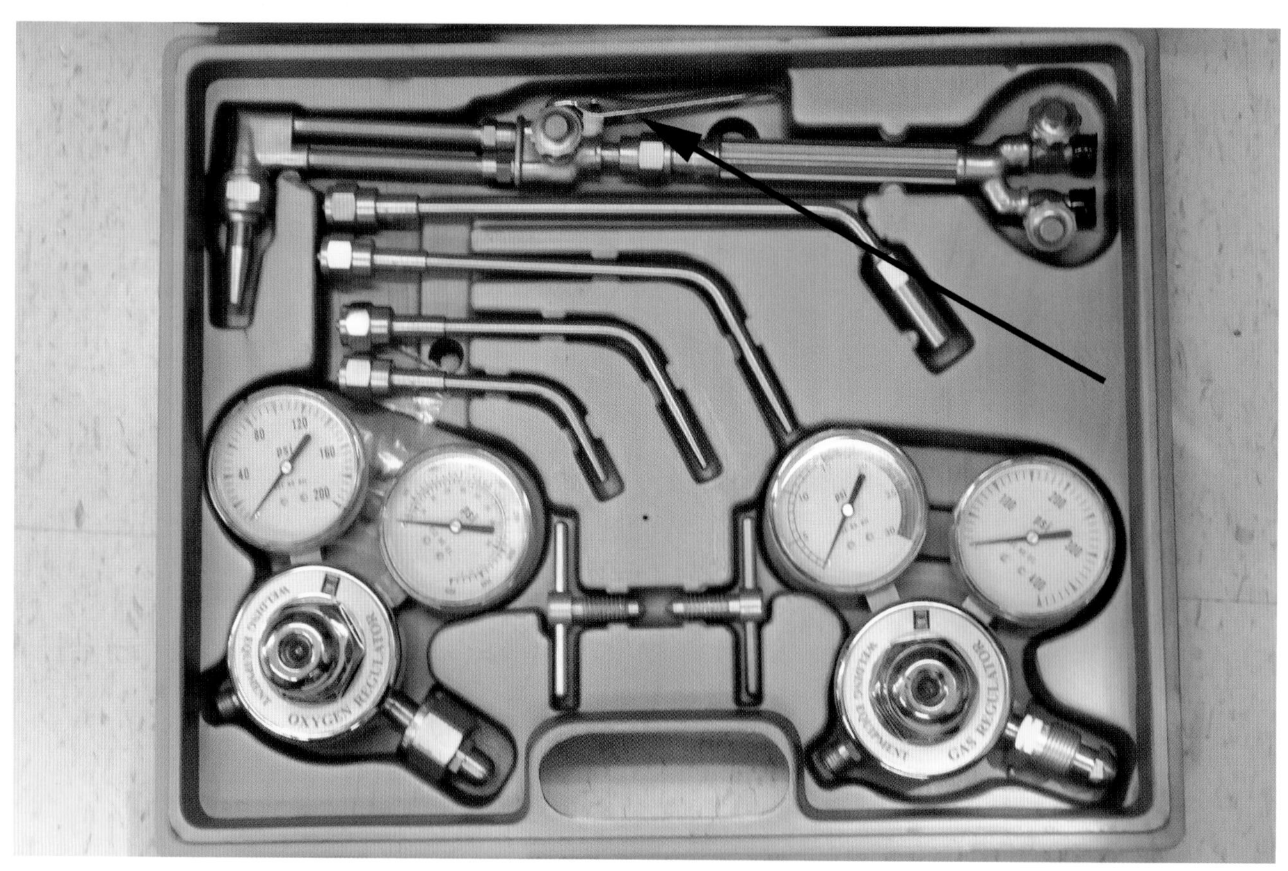

When steel is sufficiently preheated, pressing the handle indicated on the cutting torch head introduces an excess flow of oxygen that burns through the metal.

The wire-feed electrodes and gas shielded arcs of MIG and TIG welders make a lot of welding jobs easier and more practical, especially where it involves thin material or anything other than ordinary steel.

A MIG welding unit can also be used for flux-cored welding. In this mode, the weld is made using self-shielded wire with flux inside, rather than a solid wire and a gas for shielding.

- The advantages of flux-cored welding are:
- It is less affected by drafts so it's better in outdoor work.
- On rusty or dirty material it works as well as stick welding.
- Elimination of the gas bottle enhances portability.

While a good quality wire welder costs $450 to $2,000 (depending on its size), the costs for wire and gas are much less than stick electrodes. Coupled with the ability to weld aluminum and sheet metal, a wire welder can pay for itself very quickly.

The old standby stick welder is still hard to beat for many farm repair jobs where strength and economy take precedence.

TIG WELDER

TIG welding also shields an electrical arc with an externally supplied gas. Unlike MIG welding, however, it uses a different shielding gas (argon) that is inserted with respect to tungsten used as an alloying element in metals, hence the

Harmful and/or uncomfortable welding fumes and heat should be drawn away whenever possible with push-fans near the work and extractor fans at the ceiling.

name tungsten inert gas (TIG) welding. The filler material also feeds into the weld somewhat differently. Heliarc welding is an older trademarked name for TIG welding and came about from the use of helium as the shielding gas. Less expensive gases and gas mixtures are now more widely employed.

Although TIG welding is a relatively slow process, it provides high-quality welds. Typical applications are for aluminum horse trailers and irrigation pipes or stainless steel parts of food-handling equipment or sprayers. The concentrated heat and precise control of the TIG arc allows high-quality welds to be made even on thin material (0.010 inch or about 30 gauge).

A TIG welder will also be desirable if your farm workshop is also used for automotive or motorcycle hobby work. TIG welding, especially for aluminum, is the number one process chosen by professional welders for professional racing teams and by the avid auto or motorcycle enthusiast or hobbyist.

While TIG welders cost more than MIG or stick-only welders, they provide flexibility due to also having stick-welding capabilities. You'll find the machines are often referred to as TIG/stick welders. Another notable attribute of TIG welding is the very low-fume formation rate due to the filler not being required to feed through the electrical arc. TIG welding is, however, slower than stick or MIG welding, and the weld zone is difficult to shield properly in drafty environments.

PLASMA CUTTER

Plasma cutters work by sending an electric arc through a gas being forced through a constricted opening. The gas is often simply clean, dry compressed air, but can also be nitrogen, argon, or oxygen, depending on application. The process elevates the temperature of the gas to the point that it enters a fourth state of matter (plasma), which is not solid, liquid, or gas.

When the plasma arc contacts an electrically conductive material such as metal, the arc passes through the metal and melts a thin area. The force of the arc pushes the molten metal through the cut and severs the material.

Plasma cutting provides numerous advantages over other processes:

• Cuts any electrically conductive metal, including stainless steel, aluminum, brass, copper, titanium, and galvanized steel. Oxyacetylene torches only cut steel or iron.
• Fast cutting—up to 75 inches per minute on steel 3/8 inch thick.
• It does not require the metal to be preheated as with oxyacetylene cutting.
• It produces a narrow, precise kerf width (width of the cut), which is a benefit for precision cutting or when using expensive metal.
• Only a small area is heated, which reduces warping, loss of metal temper, and paint damage around the cut.
• Provides gouging and piercing capabilities.

Plasma cutters are expensive, but the smooth cut, precision, and low operating cost have made them very popular for metal fabrication, such as signs and metal sculpture.

WHAT SIZE WELDER SHOULD YOU BUY?

A decision on the size of your welder(s) is just as critical as the type of welder to buy. Duty cycle, amperage, and portability are all important factors.

Duty cycle is one way of classifying welder size. Duty cycle is the number of minutes out of a 10-minute cycle a welder can operate. For example, a MIG unit that delivers 160 amps of power at a 60 percent duty cycle can operate continuously at 160 amps for 6 minutes and then must cool down during the remaining 4 minutes of the cycle.

Amperage is another important size classification. A 130- to 150-amp MIG unit with a 30 percent duty cycle can perform many of the light repair welding jobs encountered in a farm workshop. For heavier repair or fabrication jobs, you may need a 200- to 250-amp MIG unit with a 40 to 60 percent duty cycle, or a 175- to 250-amp stick machine.

Wire-feed welders and plasmas cutters are also easily adaptable to automated computer-controlled operation, and the prices for such rigs are down to the point where they are feasible for many workshops. If your farm workshop may someday be used for small-scale manufacturing, consider welding/cutting units that are adaptable to automated operation.

THE WELDING SITE

Besides welding machines, the welding area in your farm workshop needs a few additional considerations to equip it for safe and efficient operation.

• The welding area should be located near a main equipment door so repairs can be made outside when necessary or convenient.
• The welding area should also be near the metal saw, bench grinder, and other metalworking tools so that you don't have to carry pieces a long way to be welded.
• Run a compressed air line to the welding area with either a wall-mounted pipe or a ceiling drop above the welding table so that you can use air tools, such as grinders, right at the welding bench without hoses all over the floor.
• Good lighting is essential for ease of work with tinted eye shields.
• Locate a large bucket of clean water in the welding area for quenching and cooling. Keep a light cover on it when not in use so that dirt and garbage don't fall into the water.
• To prevent sparks or slag from starting fires in the welding area, use sheet metal or fire-resistant sheathing on the lower 4 feet of walls.
• Have a large, easily accessible Category A-B-C fire extinguisher in the welding area.

When gas welding, the operator must wear protective clothing and tinted goggles. As the flame is less intense than an arc and very little UV is emitted, general-purpose tinted goggles provide sufficient protection—sunglasses are not sufficient.

For any type of electrical arc welding, a welding helmet with a dark-tinted eye shield is required. A heavy leather apron, boots, and gloves are also very useful when welding. When a molten globule of metal accidentally flies down your shirt or burns through a sock, the results are shockingly painful, and that hot little blob of metal seems to stay hot enough to burn flesh for a very long time.

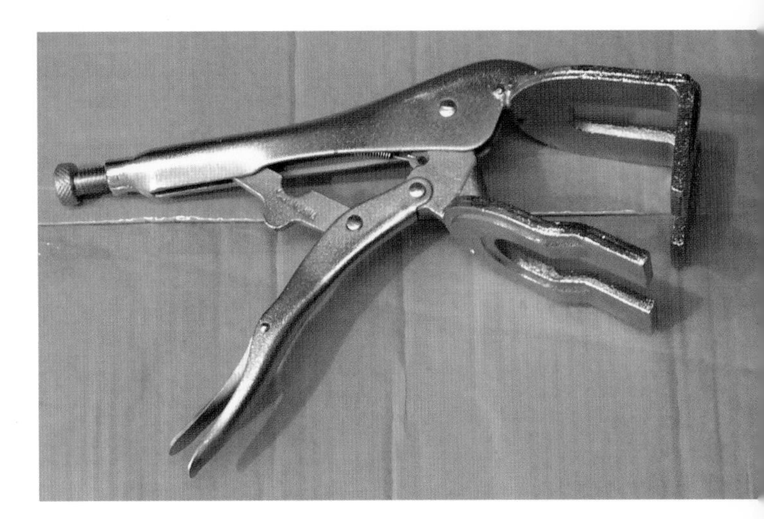

These special locking pliers provide versatile, easily repositioned clamping on welding jobs. They do tend to get metal-spattered and corroded in use, so don't count on them lasting long.

CHAPTER 5
PARTS AND SUPPLIES ORGANIZATION

Every farm workshop, big or small, needs to be equipped to deal with a wide range of repair, maintenance, and fabrication jobs. Many a simple workshop project can start off well enough but then bog down because of having to hunt a long time for the parts or supplies to finish. You may find you've collected a whole trayful of old bolts, but of course none of them fit when you need one.

Worse, the project can stall entirely and be shoved to the side as other projects become more pressing. At its worst, the project can languish so long in a disassembled pile that no one remembers why it was taken apart, what's missing, or how to put it all back together.

Those are the kinds of reasons why keeping the parts and supplies sorted so you can find them when needed is a major part of having a workshop that's an efficient and pleasant place to work.

It is a nice theory, but where does a person start making improvements? Perhaps the best place is by looking at a few situations where parts and supplies organization appears to be consistent. These situations may be instructive for other cases where organization is hard to accomplish.

EXAMPLE 1: FASTENERS

In surveying many workshops to prepare for writing this book, one system is evident in every shop: a set of bins for new bolts and nuts in various sizes and a tray holding the "odds and sods" bolts and nuts left over from various disassembly jobs. There is also usually a tray of leftover specialty

Keeping everything organized is a constant challenge in the shop, but when it is successful it pays off with being able to find the things you need when you need them.

Like things should stay alike, but the basis of likeness is up to you. When you find the way that's right for the way you think and work, it is more satisfying to work in the shop.

fasteners of the "might be useful someday" variety. Organizing principles appear to be:

Similar things are easier to organize when there are large numbers of similar ones, such as 1/4-inch x 2-inch UNC bolts that come in boxes of 10 or 100.

The system is easy to maintain when new items clearly fit existing categories. A leftover 3/8-inch nut is easy to toss into the bin holding other new 3/8-inch nuts.

The system breaks down rapidly when there are many one-off situations where the similarity isn't as obvious, such as one or two 7/16-inch brass UNF nuts. Is the category fine-thread, size 7/16-inch, brass? In practice, it's most likely to end up in the miscellaneous tray, never to be used and taking up space.

EXAMPLE 2: MOTOR FUELS

Back when farms were first adopting tractors and trucks, motor fuels were delivered in drums or metal tins and were just one of the many supplies kept in the shop. At any modern farm, big or small, tractor and truck fuels are kept in their own large tanks outside of the workshop. Organizing principles appear to be:

Things that are used often (for daily fueling) are easier to keep where they are easily accessible (in tanks you can drive right up to).

Make it easy for the holders to be refilled. In this case, situate the tanks so the fuel dealer's tanker truck has plenty of room to pull up beside them.

Keep the containers clearly labeled so diesel doesn't get put into gasoline tanks or vice versa.

Racks for front end loader attachments allow the tractor operator to drive up and attach/detach and keep clear more floor space.

Like hangers in a closet, garden tools can seem to multiply into a nasty tangle in the shop. A rack above shoulder level keeps them organized and off the floor but still within easy reach.

EXAMPLE 3: ENGINE FILTERS

Keeping oil, fuel, and air as clean as possible for long engine life is the job of the many tractor- and truck-related filters kept on the farm. In every shop surveyed, they were found grouped together near the lubricant change area. Organizing principles appear to be:

- Storage is located near where the work involving that item is to be carried out.
- Items are stored on narrow-depth shelves so the part number can be seen at a glance.

In one case, the shop owner had taken tags from the filter boxes and stapled them to the wall. This not only showed where to put the filters, but by becoming visible when the last of any line of filters had been taken for use, it indicated which filters need to be purchased to keep the shop fully stocked—*kaizen* principles at work.

During busy season farms can't afford to have machines sit idle waiting for parts so a large stock is usually kept on hand. The V-belts stored high on the wall are easy to see and reach with a long stick.

Since none of the parts for farm machinery are cheap, a well organized storage system is needed to avoid costly "can't find it" problems for the stock of parts you know you've bought.

Do these principles now stand up to the test of solving some of the more stubborn organizing problems in a farm workshop? Typical problems seem to be:

•Storing things, such as garden tools, that ideally belong somewhere else but wind up in the workshop because it's big enough to hold them.

•So many items collecting on the workbench that cleaning the bench is necessary before you can use it as a work surface.

•Stuff collecting along the edges of the shop floor where it creates a well-protected runway for mice and other destructive, dangerous, or irritating pests that sneak in through the open main doors. By the time you tear away the junk that's concealing the pest, it has escaped by running under something else.

ORGANIZING TIPS FROM THE PROS

It may not be that widely known that there is a National Association of Professional Organizers (NAPO). It was founded in 1985, has over 3,350 members worldwide, and has a Web site (www.napo.net). NAPO recently branched out to help people who are rebuilding their lives after the devastation caused by hurricanes Katrina and Rita. Six NAPO tips for successful organization are:

• Start with what's bothering you the most.

• Put prime items in prime space. In other words, keep handy what you frequently use and put the things you don't use often, more out-of-the-way.

• Tackle one small area at a time. Work on organizing one area of the shop or even one messy drawer for just five minutes each day.

• Have a helper. It's harder to toss things in a drawer while someone else is watching.

• When storing items in boxes, make sure there is a visible packing label so you don't have to open the box to see what's inside.

• When storing items for the longer term, record in two places where you put them. This can help deal with the "I bought a new one just last week, but where the heck did I put it?" syndrome.

CHAPTER 6
TOOLBOXES, CABINETS, BINS, AND RACKS

One of the first things that is often made or built for a new shop is a place to organize all the tools. This can take the form of a portable box that opens to reveal many compartments, a big rolling cabinet, pegboard or slot board mounted vertically on the wall, wall-mounted cabinets, cupboards under the workbench, or other custom-built solutions.

When you have the chance to equip a new farm workshop, any or all of these approaches may seem like a welcome change from having tools dispersed throughout a number of small toolboxes, tackle boxes, tin cans, and cardboard containers. While these containers may be a good temporary residence for tools and supplies, they aren't very professional and are easy to blame for frustration with the efficiency level of the shop.

Cupboards and wall cabinets make use of otherwise empty space under or above the workbench and are a very sensible approach. Seen in their gleaming glory in the store, the large rolling cabinets are especially attractive because of

Before you invest in any storage solution, do some observation and planning on what works in what situations.

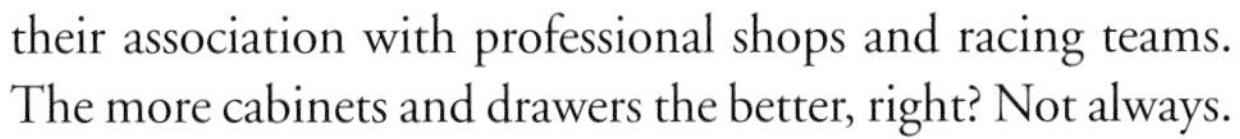

The sign posted in this shop gets right to the heart of a system where anyone can find what's needed at anytime.

When you need to handle a repair job at a far corner of your farm, it's nice to have a well-stocked toolkit you can have in the truck without stripping the shop of most of its equipment.

their association with professional shops and racing teams. The more cabinets and drawers the better, right? Not always.

In the course of observing a lot of working farm shops for this book, what became increasingly clear is that each of the tool storage solutions may have its place in a shop, and each can just as easily prove to fail as the desired solution. The most common situation is that the top few drawers of large rolling cabinets are well organized, the contents of the middle drawers become noticeably more chaotic, and the bottom drawers and shelves are a real jumble. Another common situation is that cupboards under the workbench become little more than collection points for rarely used junk. Wall-mounted racks seem to be reasonably well organized at one end, while the other end looks strangely empty. This indicates that its capacity as an organizing tool seems to fade quickly.

To get an answer to why solutions that work in a professional garage, industrial setting, or hobby space don't work as well in a farm workshop, it's worth considering the differences in what the shops do. For example, truck or tractor dealer shops are dedicated to working on particular brands, types, and years of vehicles, and a fairly narrow range of jobs are most often required. The mechanic is familiar with the range of tools needed and keeps them in a cabinet that can be locked up at the end of the day to prevent theft or borrowing by others. This is taken to an extreme by racing teams who only work on one vehicle and have to do a few jobs really fast. Any unusual repair probably requires so much time that they have to withdraw from the race. Even a shop dedicated to a hobby, such as boat building or woodworking, has a reasonably well-delimited set of tasks involved.

Contrast this with a farm shop, where over the course of a week you may need to weld a cultivator frame, change the oil in a truck, track down electrical faults in a tractor's instruments, change seals on an irrigation pump, install a stereo on the daughter's car, calibrate planter units, replace rotted boards in a manure spreader floor, rebuild a carburetor on a vintage tractor, and so on with near-endless variation.

The big difference is that in a farm workshop you tend to do a lot of different kinds of work on a lot of different kinds and vintages of machines. The work is anything but repetitive, and that kind of variety is something many farmers enjoy. But it does complicate the problem of how to achieve an efficient arrangement of tools. How can one arrange the tools if you're never really that sure what needs to be repaired from day to day or hour to hour? It's not like a commercial shop where you can look up the job in the flat-rate manual and lay out, in order, the exact tools you'll need.

One solution lies in being able to easily rearrange the tools to suit the particular job. This approach is being used more often in modern factories in order to quickly change to whatever customers need built. Instead of having large, hard-to-move, efficient-for-a-single-purpose equipment ("monuments," as they are called in the literature about industrial processes and efficiency), modern factories try to have equipment and assembly areas that can be rapidly reconfigured to suit the job, avoid downtime, and keep overall productivity higher.

The way this can be applied to the choice of tool storage in the farm workshop is by having storage systems that can quickly be reconfigured to suit the job and avoid large monuments where you have to bring the work to them. For example, having tools hung on pegboard or nails

More farms are being equipped with mobile service trucks that are a complete workshop on wheels.

Large pegboard racks seem like such a good idea that it's too bad they rarely work in practice.

above the workbench can create a monument that's hard to deal with for years to come. It seems like a great organizational idea, but in practice it may mean you have to bring all work to the bench (not always practical or possible) or disrupt your workflow by walking over to the bench to get a needed item. In contrast, having a wide range of tools in some sort of easily moveable storage lets you bring the tools to more or less exactly where you need to work and produces less disruption in the workflow.

The reduced disruption is important in a farm workshop because it reduces stress for you. Farm repair jobs often already involve the stress of trying to get a piece of equipment back into operation while favorable weather and/or crop conditions hold. There is no point in adding stress by having to run back and forth to the bench to get a tool you discover is needed and can't be found because it wasn't put away after someone used it.

What complicates the issue again is that the very same tool storage device (tools hung on pegboard or nails above the workbench) can be a useful organizational tactic in some areas of the shop. Woodworking or small engine repair are two examples. In both cases, much of the work is done on the workbench in a relatively compact space, so it makes good sense to have all the tools easily reachable in that area. Where work must be taken off the bench, it can go to monuments, such as the woodworker's table saw or the engine mechanic's bearing press where ordinary hand tools are not needed.

Where similar workflow characteristics exist, such as in a lube bay, fixed tool storage can promote efficiency. At these points in the farm workshop, it may be worth having a duplicate set of the required tools kept at the site where the work always occurs. These days a decent quality set of combination wrenches or a socket set is cheap enough to make the extra-tool approach a very realistic option. In other areas, such as where an unpredictably wide range of repairs is made in the middle of the shop floor, it makes more sense to use moveable, reconfigurable tool storage.

Configuring tool storage to meet arising needs can be as simple as using a carpenter's tool belt, apron, or bucket-mounted tool bag. Rolling a tool cabinet over to the job can be handy, but it often seems to end up on the opposite side of where you're working and you have to stop and walk over to it. Another useful approach is a small rolling trolley that can carry a fairly broad range of tools (e.g., wrenches and sockets) in the top tray and more specialized tools (e.g., a multimeter) and specific repair parts in a lower tray. If the work area will be low enough, a rolling seat with tool tray underneath combines convenience and a comfortable working position.

KEY POINTS:

- Avoid having all your tools in a fixed place (e.g., racked above a workbench) unless you can be sure the work is going to be done only at that place.
- Look for moveable tool storage solutions, such as tool bags, small trolleys, or rolling seats with a tray underneath.

Cabinets help get clutter off the bench but often wind up transferring the clutter behind closed doors. It is not much help when you need to find something.

In many cases, the tools will have to go out to the field in order to service a breakdown or because it will take too much time to bring the machine to the shop. If all the tools are centrally organized in one container or rack, a field repair may leave the workshop short of a necessary item. The current cost of good-quality tools is low enough that it may be worth having several duplicate sets, either in separate field toolboxes or a dedicated service truck.

PUTTING THINGS BACK

No matter how good your tool storage is, it's not going to work for very long if things are never returned after use. Many farm workshop owners express frustration over how messy their shop gets and how their children or employees show no interest or ability to replace tools or replenish supplies that have run out in the course of a job. Yet some of the farm shops photographed for this book were astonishingly well ordered and clean, apparently all the time. Why would there be such a striking difference? There is no doubt that some of it depends on personal habits. Even where the shops were "organizationally challenged," none of the farm areas outside the shop displayed any lack of organization, nor were there heaps of things that were going to put away at some undefined time in the future. It did not appear that any of the shop owners were overall messier than the others.

More of the difference may depend on why the tools were taken up in the first place. Getting the tools out in the first place solved various problems as they came up. For most, solving problems is much more interesting than cleaning up afterwards. Putting the tools back may now seem like a minor problem that's too easily solved by leaving it for another time. The far more pressing problem, in many cases, is getting the repaired piece of farm equipment back into operation—before the rain and in time for planting, spraying, or harvesting. Not so much a lack of dedication of order as a concentration on solving the big problems first may be the crux of why things don't get put back.

Rather than changing that focus, it might be easier to make the smaller unsolved problems easier to solve so that they may cease to be problems at all. As a rule of thumb, the easier it is to access the tool in the first place, the easier it is to put it back when you're done. If getting the right tool involved walking over to a fixed workbench rack or other monument type of storage, it's that much less likely that you'll walk back to put it away until the job is definitely over, and by then some other demand on your time may have arisen and cleaning up doesn't happen.

In contrast, if at the end of the job you can toss all the tools on a tray or trolley and carry that one item back to the original storage area, it becomes a less daunting problem to put everything back.

That is how smaller, more flexible storage arrangements can help solve the problem of tools being returned to their proper homes, or at least keep the problem from getting completely out of hand. Instead of dreaming about a big silver bullet approach to tool storage in the farm workshop, it may be better to adapt the type of tool storage to the many different types of work that need to be carried out.

Note how many of the tools in this shop are on wheels so that they can be moved easily to where the work is, not the other way around. That's an underlying principle of effective tool storage.

CHAPTER 7
WORKBENCHES

A key element in any farm shop is arranging tools and supplies for repairs, assembly, fabrication, or modification. Needs include:

- A stable, well-lit, accessible platform to support and hold materials.
- A convenient height in order to make your work more efficient and prevent repetitive motion strain.
- Tool access that makes tools easy to find and put away.
- Easy access to electrical outlets for power tools.

The height of the work surface is a key component in whether the bench is comfortable and practical to use. A rule of thumb is that for general purposes, bench height should be about at wrist level, which of course varies according to your height. For precision work such as working on a tractor carburetor, the bench can be a bit higher so fine details are easier to see. This can be achieved by having a large box of suitable height that you can set on the regular bench top to bring precision work up a little closer to the eyes.

Workbenches are usually recognized as one of the first needs in a shop, and a good design will help them stay useful for years to come.

Part of the workbench design is keeping the cleanup of benches from becoming a major recurring chore.

Commercial workbenches heights are in the range of 33 to 36 inches, which fits the needs of most people. If you're buying a bench, shop around to check for the height that feels best to you. If you need a lower bench height, the easiest way to change things may be to raise yourself by putting a thick rubber mat in front of the bench, which also increases comfort while standing. If you're taller and need a higher bench, you can put it on sturdy blocks or caster wheels.

DESIGN BEFORE YOU BUILD

A key problem with deciding on an efficient workbench in a new farm shop is that you often don't exactly know what type of work the bench will have to handle, so it's hard to come up with a suitable design.

The design solution suggested in this chapter is for an efficient, satisfying workbench setup in a farm shop, the best approach may be to start off with a movable, modular workbench design that can adapt to the needs and workflow that emerges over time in your own situation.

To some extent, the applicability of this proposition depends on your own work and organization style. If you are very organized and always take time to put tools away in the right place after use, any bench system may work just fine. But if keeping tools organized and bench tops clear of clutter tends to prove a problem (and you're certainly not alone if it is), a nontraditional system might prove more efficient in your shop.

A traditional approach in farm shops is to build 3-foot-high workbenches against the walls. To achieve a stable work surface, attach the benches firmly to the wall. The vertical surface at the rear of the bench may be utilized to hang tools from hooks or nails for at-a-glance storage.

In use, however, it has been the author's observation that in an awful lot of farm shops, these benches often wind up being cluttered junk magnets for two reasons: they do not make it easy to put tools away after use and the large open surface is a convenient place to pile things such as spare parts, papers, and clothing that do not belong there.

Shelves under the bench may just shift storage problems out of sight and not really solve them.

Commercially made workbenches already have the advantage of being able to move around, but as they get filled with tools, moving the bench becomes harder.

An early example of a specialized workshop bench—the anvil. It combines surfaces for hammering, bending, punching, and tool-holding, and is mounted on a pedestal so it can be accessed from any side.

The upshot is that someone coming to the bench to work on a task has a hard time finding tools, and cleaning up the cluttered bench top becomes a long, dreaded task in itself—not exactly a desired improvement in efficiency! In practice, it has also been observed over the years in many farm shops that traditional drawers for tool storage under the bench tends to accumulate the least-used tools. It's too easy for the out-of-sight, out-of-mind approach to cleanup to take over. Natural workflow seems to wind up placing the most useful tools elsewhere.

In terms of productivity and convenience in use, since the side attached to the wall is inaccessible, the work itself must be turned around if access to the other side is needed. That can be quite inconvenient if the item being worked on is not exactly stable or may go out-of-square when being moved before completely assembled. On a long bench the operator may have to make awkward bends to reach the ends of the work piece.

To get past the clutter problems that seem to go along in practice with traditional designs, another approach is to have workbenches that are:

- Smaller so they can only accumulate the things you are presently working on.
- Modular so that several can be clamped together if you need a larger work surface.
- Movable so workroom is available all around the bench. Alternatively they can be moved right next to where you happen to be working, such as beside an implement.
- Specialized to suit particular tasks.
- Adjustable in height to accommodate various projects.
- Folding workbenches with clamping capability, such as Black & Decker's Workmate benches, are one way to achieve many of these goals. Once you see where and how you use them for various tasks, you can start realistic planning of larger, permanent shop-built benches. Some models are available with built-in electrical power bars.

Another example of a home-built alternative workbench of a large size is the Experimental Aircraft Association (EAA) Chapter 1000 Standardized Work Table. These tables have tops 2 feet long by 5 feet wide and

A rolling tool cart lets you bring all the necessary tools to the job, but it also allows you to load everything into one convenient place when the work is finished, which makes it a lot easier to put tools away.

The mechanic's rolling stool increases efficiency by combining a simple centralized tool tray with a comfortable seating platform.

are 33.75 inches high. Vises or jigs can be attached at convenient places. When needed, several tables can be clamped together to make a work surface suitable for long or wide pieces. You can even slide the benches back against the walls if that becomes appropriate for a particular job. Height can be adjusted during construction to suit your body dimensions and preference. For electrical needs, power bars can be attached at any convenient place so that only one cord needs to run from a wall outlet to the bench. Detailed information and plans for these easily constructed benches can be viewed online at: www.eaa1000.av.org/technicl/worktabl/worktabl.htm. For information on paper plans, write to EAA Chapter 1000, 3435 Desert Cloud Ave., Rosamond, CA 93560-7692.

For working on medium-sized equipment, such as lawn mowers or small imple-

A vise is often handy to have in places other than bolted to a bench. This pedestal-mounted unit allows placement of a vise exactly where it's needed most.

ments, a modular worktable and chain hoist can be put to use. Position the item under your shop's chain hoist (refer to Chapter 8) and suitably strong chains or non-marking tow straps, lift it high enough to allow the worktable to be slid underneath, and lower the item to a secure working position on the table.

SPECIALIZED BENCH EXAMPLE: THE WELDING BENCH

In some tasks, such as welding, there are a limited number of particular needs that are well understood. Welding bench designs take those needs into consideration in order to make welding efficient and easier on the operator.

The heavy steel construction of the welding bench permits hammering and chipping, and also improves grounding conductivity when electric welding. Because the welding bench is freestanding, operators can move all around the bench to make welds where needed. This allows complex pieces to stay clamped in the vise or a tabletop jig until all welds are completed.

STOOLS AND CREEPERS

Sometimes work on larger machines, such as a tractor, can't be put on a convenient bench. In this case, putting yourself on a special work platform improves efficiency and working comfort.

For working pieces and equipment that would ordinarily be at about waist height and require stooping, a mechanic's rolling stool provides a comfortable perch. The tray under the seat provides storage for the tools you need to use, and all the tools remain visible.

A newer creeper design addresses some of the problems associated with conventional small-wheeled units.

For working under vehicles, a mechanic's rolling creeper allows you to roll to the necessary places without lying on cold, wet, or dirty floors. If you can find the type with an attached side tray, it is a lot easier to bring the necessary tools along with you as you slide under the vehicle. Be sure to wear eye protection because dirt, rocks, and various fluids may drip down on to your face.

FOR MORE INFORMATION ON WORKBENCHES

For a general-purpose workbench, good places to start looking are the many magazines and books on woodworking. One example is *How to Design and Build Your Ideal Woodshop: Revised Edition* by Bill Stankus. Many plans and ideas are available online, such as the EAA Chapter

Innovative design makes it possible to use this creeper while working on heavy equipment outdoors on gravel, which is something that's impossible with older designs.

Wheel size is the key difference in creepers. The larger wheels roll easily over small bits of gravel or floor cracks that would stop conventional small wheels. The larger wheels necessitate the arches in the body that give this creeper its bone shape, which is also more comfortable for the mechanic.

1000 modular bench mentioned above. To see many more examples, do an Internet search for "build workbench."

Welding table plans are available in many books, including *The Welder's Bible, 2nd Edition* by Don Geary. Plans are also available online at sites such as www.progressivefarmer.com/farmer/projects/article/0,24672,1087277,00.html. Do an Internet search for "build welding table" to see many more Web sites to choose from.

Plenty of examples of commercial workbenches can be seen at hardware suppliers, or online at sites such as www.waterlooindustries.com and click on the workbenches section; and www.blackanddecker.com and go to the power tools section and then under the Workmates category.

CHAPTER 8
LIFTS, HOISTS, JACKS, AND PRESSES

Many jobs in the farm workshop will involve lifting and moving heavy pieces of material, which are often awkward and involve twisting the body and extending the arms to get the piece out of a machine or back into position. For example, removing the cylinder head while rebuilding a tractor engine involves lifting a hefty lump of cast iron and turning to place it on a bench or cart.

Lifting and hoisting tools can often make this kind of work considerably easier. They are not always a solution because sometimes there is simply no easy way to hook them to the work or make the lift. For example, removing and replacing the heavy (50 to 100 pounds) concaves, chaffer, and sieve is a typical job in a combine overhaul, and may even need to be done several times per harvest to clean out a combine used to harvest seed. But use a lift whenever it's practical, especially where you can reduce the risk of damage to parts and muscle strain or back injury for the mechanic and assistants.

When hoisting objects into the air is required in your workshop, one thing that is not a solution is a winch. Winches are meant for pulling objects that do not lose contact with the ground or with some other friction-creating

Along with lifting equipment for service, your workshop needs the often-overlooked means to securely hold the equipment in the air while you're underneath it.

A machine that accidentally drops to the floor while it is being worked on can result in costly damage to expensive parts.

Wooden blocks, especially tough hardwoods, make one of the best ways to support heavy equipment securely. Note how many short pieces have been cribbed together to make a very strong, larger support.

surface. Their purpose does not include pulling things up and holding them there in the absence of friction. The design of a winch is such that if you used it to lift a heavy object, free spooling could bring that heavy object crashing down with potentially disastrous results all around. The same sorts of problems occur with using a rope-and-pulley lift. There is nothing to stop the load from falling rapidly if the rope slips or breaks.

One of the most versatile solutions is the chain hoist. Moving the smaller chain rapidly with light hand pressure slowly raises or lowers the larger lifting chain. Ratcheting and electrically powered models are also available. Two valuable advantages of the chain hoist are that it is very low-effort and when you stop lifting, the work hangs securely in place to make alignment easy during reassembly.

The chain hoist can be attached to any strong point, such as a roof beam, above the object to be lifted. Just be sure the strong point can bear the weight you'll be lifting, or the roof might come down instead of the piece going up! The chain hoist can also be attached to a rolling gantry so you can roll the whole lifting apparatus to where you need it, and then roll the lifted piece over to where you want to work on it.

Some chain hoists are made to mount on a moveable trolley system. This allows the load, once lifted clear, to be pulled sideways. By using double trolley rails mounted at right angles, you can also move the load forward and back, achieving maximum lift-and-move versatility.

Attaching the chain hoist to a tractor's raised front-end loader bucket creates several hazards and should be avoided if at all possible. If you must use the front-end loader, observe sensible precautions, such as:

1. Make sure the rear end of the tractor (without operator in place) is heavy enough to counterbalance the weight being lifted. If required, add temporary weight at the rear of the tractor with a ballast box or heavy implement on the three-point hitch.

2. Attach the hoist very securely to the front-end loader with a suitably heavy piece of chain. Hooking a chain to the lip of the bucket may allow the hoist to slip sideways if it becomes slightly off-balance. Lock the hook securely in place with clamps or locking pliers.

3. Mechanically brace the loader lift arms to prevent them drifting downward. Do not count on hydraulic power alone to hold the arms up, whether the engine is running or not. One way to brace the arms is to slip a suitable length of heavy channel iron over the exposed portion of the loader lift ram, and then secure the iron in place with straps or several windings of strong rope.

4. Do the final lifting with the hoist, not the loader. Front-end loader controls are rarely smooth enough for this kind of work.

THE SHADE TREE MECHANIC

. . . A handful of tools, a vise, a hoist, and a stout lower limb on a white oak tree were all that were needed to maintain either tractor or truck. An 8- or 10-inch white oak limb was sturdy enough for lifting the front end of a vehicle or for hoisting the engine out of its mounts. In the summer, the canopy provided shade. Although their skills would eventually be derided, "shade tree mechanics" were the familiar mechanics of the countryside, and they collectively embodied a set of skills and values that [were] defining features of self-reliance in twentieth century America.

from *People and the Land: Elegy, Memory, Promise* by Gerald L. Smith and Andrew Gallian

Chain hoists use mechanical advantage to put tremendous lifting force in easy reach. The overhead trolley rails and rollers make it easy to position a hoist in your shop.

Another approach to a trolley hoist is to put it on a stand that can be rolled to wherever it might be needed inside the shop or out in the yard.

This specialized engine hoist is designed so that the lower support slips under the front of the vehicle to bear the weight of the hoisting operation.

If the piece you intend to hoist is large, awkwardly shaped, and/or has no easy way to hook up the lifting chain hook, lifting tabs and load-spreader bars can make the job easier and more efficient. Lifting tabs can be fabricated in the shop using suitably sturdy pieces of flat steel strap. Drill a suitably sized hole in one end to allow the tab to be bolted to the work piece, using one of the bolts that is already part of the machine. If necessary, the existing bolt can be removed and a slightly longer one can be threaded in for attaching the lifting tab.

In the other end of the tab, drill a hole big enough to attach the hook of the lifting chain or bolt the chain directly to the tab. In some cases, it may be handy to heat the tab and give it a twist so the tab can lay flat against the work piece and the hoisting hook can still be easily attached. If you do heat the lifting tab, quench it in water to preserve the strength of the metal (see the Tempering and Annealing sidebar in Chapter 4).

Spreader bars can also be easily fabricated in the shop to help with achieving straight vertical lifts on a piece with multiple attachment points. Select a piece of square or round metal tubing long enough to extend over each available lifting point. If there are more than two points, tubing can be welded up into a frame. Drill holes in the ends of the arms so you can attach short lengths of suitably heavy chain that connect to the lifting points with hooks or bolts. Then connect the lifting chain of the hoist to a point where it lifts the spreader bar evenly.

Some experimentation may be needed to find the point where the hoist spreads the lifting force evenly to all the short chains at the ends of the arms. If you raise the hoist and one of the short chains remains loose, move the central lifting point toward that chain. Eventually you will find a point where all chains are evenly tensioned. But before you go ahead with the lift, make sure the central hoisting chain is securely attached to the spreader bar. If it slips during the lift, the work piece may suddenly lurch to one side and damage the work piece, the machine underneath, or any fingers and arms in the area.

The basic mechanical jack is useful in many places around the farm. It is prone to balky operation as sliding pieces dry out or become corroded, so keep it well lubricated, preferably with a dry-type lubricant that does not attract dust.

Hydraulic lifts that move an entire machine vertically upwards are not found as often in farm workshops. They are useful for smaller items, such as when you need to work on a lawn tractor or garden tiller or for working on hobbies such

The hydraulic press is capable of exerting tremendous amounts of force in precisely controlled increments. Pins in the side rails permit positioning the lower jaw at a suitable height.

Simple ramps are useful to simultaneously lift and securely support vehicles that need service. Knowing when to stop as you drive on the ramp is one of the perils of using a ramp, so place blocks ahead of the rear wheels to give you an indication of when the front wheels are on the flat upper part of the ramp.

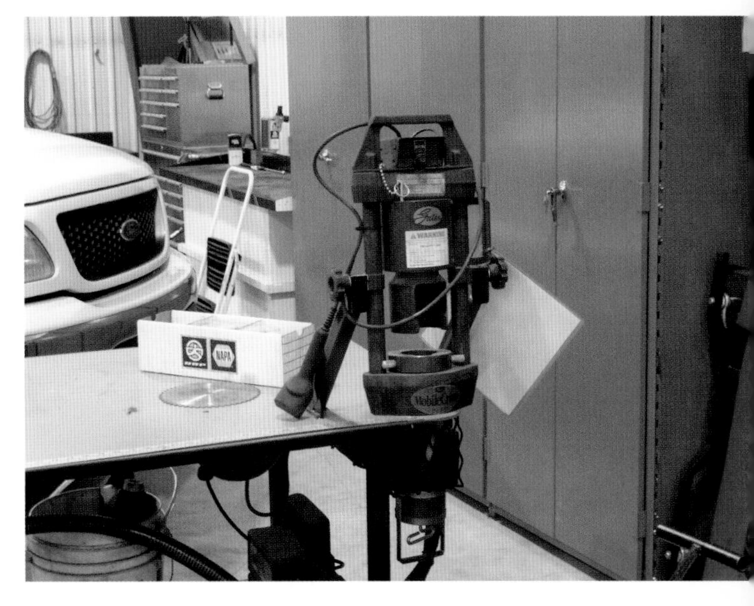

The hydraulic fitting crimper is a special type of press that enables hose repair and custom fabrication in the farm workshop.

as cars, ATVs, or motorcycles. But the tractors, implements, and large trucks that account for most farm workshop jobs are tall enough to be worked on without lifting and are too heavy to lift without impracticably large machinery.

JACKS AND SUPPORTS

Machines do occasionally need to be partially lifted off the ground for work such as changing tires or servicing axles. Jacks and supports are needed for this kind of work; jacks to do the lifting and supports to hold the machine in place. The certainty of a separate mechanical support is necessary after lifting. Do not rely on a jack alone to hold things up—many farmers have been killed or injured by making this mistake. A stout wooden block or section of log is often enough, and a well-equipped shop will have plenty around to suit whatever needs arise.

Jacks are also useful for horizontal spreading of parts, such as when the spacing between tillage assemblies is adjusted by sliding them along the implement's toolbar. Ideally one could loosen the clamp and slide the assemblies using hand pressure and a little oil, but sometimes they just won't move that easily, and prying on them with a bar is awkward or there's no place for good leverage. A small hydraulic bottle jack helps move them slowly to the correct point along the toolbar. However, when resorting to a jack to move stubborn parts that won't separate by pushing or tapping, be careful about several things:

Check and check again that there are no bolts, pins, or clips still fastening the parts together.

Don't apply so much force with a jack that you break the part being removed.

As you apply pressure, keep an eye on the base of the jack so that you can make sure it does not break or bend the part it is braced against.

Arbor presses are available in a range of sizes and have specialized ends to permit various types of work.

More tools, such as this floor jack, are using compressed air to actuate hydraulics for easier, faster lifting jobs.

This large hydraulic bottle jack is actuated by compressed air. In use, there is often limited room to operate the pumping handle of an ordinary bottle jack so pneumatically activated ones are quite handy.

Apply force slowly and evenly, keeping a close eye on the effects, and while you're slowly jacking things apart, tap, wiggle, and lubricate the pieces to help free them.

PRESSES

Along with things that need to be lifted up, some farm workshop jobs will involve things that need to be pressed down with quite a lot of force. For example, a seal may need to be pressed into a cover, a shaft may need to be pressed out of its holder, or a bent piece of metal may need to be pressed straight again. For these jobs, two types of presses will be useful.

The arbor press uses a rack-and-pinion mechanism to push the moving shaft against the base. The rugged body casting keeps the whole thing from twisting when force is applied. Since hand pressure is all that's used with the arbor press, fine control of the shaft movement is possible. Pressure in the range of 1 to 5 tons is possible with these types of muscle-powered presses.

The hydraulic shop press exerts much more massive amounts of force, typically in the range of 30 to 50 tons. These presses can also be equipped with kits to punch holes in metal up to 1/2-inch thick. The advantage of a hydraulic shop press for straightening metal is that heating is not required. When metal is heated and cooled, changes occur in its strength and hardness (for more complete details, see the Tempering and Annealing sidebar in Chapter 4). By using a press to bend the metal back into shape, these changes are avoided and the metal stays at its original ductility and hardness specification.

The bent piece might happen to break in the process of pressing it straight, especially if a lot of force is applied quickly. But chances are if the metal bent and did not snap when force was originally applied, it will also bend back to its original shape when a similar force is slowly applied in an opposite direction by the hydraulic press. By arranging various blocks and pieces of metal as spacers between the work piece and the fixed lower plate of the press, you should be able to position the work piece so the ram presses directly down on the area to be bent.

The rivet press is a similar type of muscle-powered press used when replacing individual cutting teeth (sections) on sickle mowers, combines, or swathers. The new rivets are inserted from below, then screwed tight so that the pin of the rivet press contacts the top of the rivet while the bottom is held firmly by the body of the press. As the pin is tightened with a wrench, the top of the rivet mushrooms out to lock it in place. Rivet presses are sold at equipment dealers and farm supply stores and should be carried along in the toolbox whenever you are using a sicklebar cutter. New sections, plus a hammer and chisel to remove the old rivets, will also be needed.

CHAPTER 9
PARTS CLEANING AND DEGREASING

For many jobs in the farm workshop, it will be necessary or desirable to clean parts either before or during the work. Getting the dust, mud, chaff, and manure cleaned off tractors and implements before they ever get into the shop makes the work a lot more pleasant and removes contaminants that could make their way into engines and transmissions during repairs. Grease, sludge, and scale will need to be removed from engine and driveline parts such as drive shafts, oil pans, and cylinder heads in order to assess component condition and restore full performance capabilities. Corrosion and paint will need to be removed when you're welding or touching up the paint or restoring old machinery. For bearings that need to be repacked with grease, it's usually recommended that you wash the old grease out first, as it may have lost a certain amount of lubricating quality or picked up various levels of dust and other contaminants.

The workshop system to accomplish these cleaning tasks involves several specialized types of cleaners. Each has a specific range of use and particular characteristics that influence its place in the shop.

PRESSURE WASHER

A pressure washer is the first step in preparing an item for subsequent operations in the workshop. Pressure washers use an engine and pump to boost water pressure and volume above what's supplied by the intake hose, as well as adding heat and chemicals to clean very quickly and thoroughly. Pressure washers also improve water use efficiency: an average

A pressure washer makes quick work of cleaning off machines before they bring mud, manure, and straw into your farm workshop.

This is a cabinet-type sandblaster that shows the hand ports that allow manipulation of the part being cleaned. A clear window in the top of the cabinet allows you to monitor the cleaning process.

pressure washer uses approximately three to four gallons per minute (GPM) of water, whereas the average garden hose puts out somewhere between six and eight GPM.

Unless the shop has an indoor wash bay with floor drains and spray-stopping curtains, the pressure washer will be used most often outside the shop. The pressure washer needs to be connected to a water source (and electricity, for electric models), so the workshop should have outdoor outlets to eliminate the need to string water hoses and electric extension cords long distances. For larger cleaning jobs such as farm equipment, pressure washers equipped with gasoline engine–powered pumps are also available. If frequent tough cleaning jobs are expected in your workshop (e.g., frequently cleaning out livestock trailers), pressure washers and steam cleaners are also available as oil-fired and natural gas–fired models.

• Cold pressure washers used with or without soap are usually sufficient for dirt, dust, mud, manure, and other cleaning jobs where grease and oil are not a concern.
• Hot pressure washers are good for quickly cleaning large areas covered with light to medium oil and grease but where the grime is not caked on or in a very thick layer. The hot water helps loosen oily grime and improves the performance of the soap in the spray. Hot pressure washers will leave a slight grease or oil residue.
• Steam cleaners will clean more thoroughly than a hot pressure washer, but will take longer on large areas. The high-temperature steam will not only clean oil and grease, but also melt down such substances as honey, tar, and many types of glue. Steam heat also makes soap more aggressive. Steam cleaners have excellent sanitizing capabilities, so they are a good choice where you need to clean off food-handling equipment.
• Combination units are available to handle any type of cleaning.

This is a cabinet-type sandblaster that shows the hand ports that allow manipulation of the part being cleaned. A clear window in the top of the cabinet allows you to monitor the cleaning process.

Pressure and volume ratings give an overall indication of how quickly a pressure washer can work and influence the cost of the pressure washer. Pressure ratings themselves are a guide to how deeply the unit can clean. High pressure in the cleaning fluid stream helps break the bond holding grime to the surface being cleaned. Pressure ratings range from 750 to 5,000 pounds per square inch (PSI). Pressures above that level may start to cause damage to the surface being cleaned and atomize the spray too finely. If the stream of cleaning fluid becomes mist, particles lose velocity too fast and cleaning becomes ineffective.

A small gravity-feed sandblasting gun can be used for detail work and touchup after using a large unit. Protective gear, especially eye protection, is required when using this type of sandblaster.

Volume is a guide to how fast a surface can be cleaned because volume of cleaning fluid is what flushes the dirt away once the bond to the surface has been broken. In general, the higher the horsepower rating of the pressure washer, the higher pressure and/or flow it can deliver. For washing down farm equipment, a recommended minimum pressure-washer rating is 4 gpm at 3,000 psi.

Other factors in choosing a pressure washer include whether you intend the unit to be portable or stationary, whether it is electrically powered or has its own internal-combustion engine, and how often you expect to use the machine. A rule of thumb is:

Light-duty, consumer-grade pressure washers should be sufficient for up to 3 hours use per week. These units typically use less-costly, disposable, axial-type pumps.

Medium-duty commercial/agricultural-grade pressure washers can handle up to 20 hours use per week. These units typically upgrade to a rebuildable triplex plunger–type pump, which often has double the life of an axial pump.

Heavy-duty industrial/commercial-grade pressure washers are needed if the expected use is more than 20 hours per week. For longer life in tougher operating conditions, rebuildable triplex plunger–type pumps in heavy-duty pressure washers may have durability and performance features such as higher-grade bearings, oversized connecting rods, polished solid ceramic plungers, and more.

Along with the basic equipment that comes with the pressure washer unit, your workshop should also be equipped with a selection of spray nozzles for various cleaning conditions. The degree rating of the nozzle refers to how far the spray fans out as it emerges from the nozzle.

Nozzle type: 0 degree (blasting)
Typical uses:
- Removing caked-on mud from farm equipment.
- Cleaning tar, glue, or stubborn stains from concrete.
- Cleaning overhead areas.
- Removing rust from steel and oxidation from aluminum.

Nozzle type: 15 degree (stripping)
Typical uses:
- Removing paint from metal or wood.
- Removing grease or dirt from equipment.
- Removing rust from steel and oxidation from aluminum.

Nozzle type: 25 degree (cleaning)
Typical uses:
- General cleaning of dirt, mud, and grime.
- Rinsing surfaces in preparation for painting.

Nozzle type: 40 degree (washing)
Typical uses:
- Light cleaning and washing.

The place where you buy your pressure washer will also have a selection of soaps and other cleaning chemicals to use in various situations. These products are usually sold in a concentrated form and need to be diluted with water to the correct level before use. Refer to the pressure-washer operator's manual for guidelines on cleaning products to use and how much they should be diluted.

SANDBLASTING

Sandblasting is a fast way to remove paint, corrosion, rust, and other hard contaminants from parts. Sandblasting works by using compressed air to bombard a part with parti-

This bench-top solvent parts cleaner has a recirculating pump and gooseneck hose to continually bathe dirty parts in solvent.

cles of abrasive materials. The abrasive material used may be ordinary sand, aluminum oxide, glass beads, broken walnut shells, or other materials that suit the part being cleaned.

Pressure-feed or siphon-feed sandblasters are used outdoors due to the high volumes of dust involved. Both types require 80 to 125 psi of compressed air. Siphon-feed sandblasters operate by drawing abrasive from an open container through the feed line. This type of sandblaster is less expensive and physically smaller, but it has the disadvantage of using part of the compressed air energy to draw abrasive up to the outlet. This results in reduced blast energy compared to a pressure-type sandblaster.

With pressure-feed sandblasters, the abrasive material is poured into a tank that is sealed and connected to the compressed air supply. The air-abrasive mixture comes out with much greater force than from the siphon-type sandblaster. For the same cleaning job, a pressure-type sandblaster typically uses 1/3 the abrasive and considerably less air volume than a siphon-type unit.

Because sandblasters are so effective at stripping parts down to bare metal, be ready to apply some sort of anticorrosion layer (e.g., a light spray of wax or oil) once the sandblasting is complete; otherwise the newly exposed metal will quickly rust or corrode once exposed to oxygen and moisture in the air.

SOLVENT WASHING AND DEGREASING

Once the equipment has been cleaned off enough for closer work, oily or greasy parts may need to be cleaned off afterwards with solvents. A stiff brush and a panful of low flashpoint solvent may be enough for occasional small jobs, but dedicated parts washers have the advantages of being more convenient and less wasteful of an expensive solvent. Current prices of automatic parts washers are low enough to be well within the range for any farm workshop. Some large models also feature a safety link in the lid hinge so that if there is an accidental fire, the lid automatically closes to smother the flames.

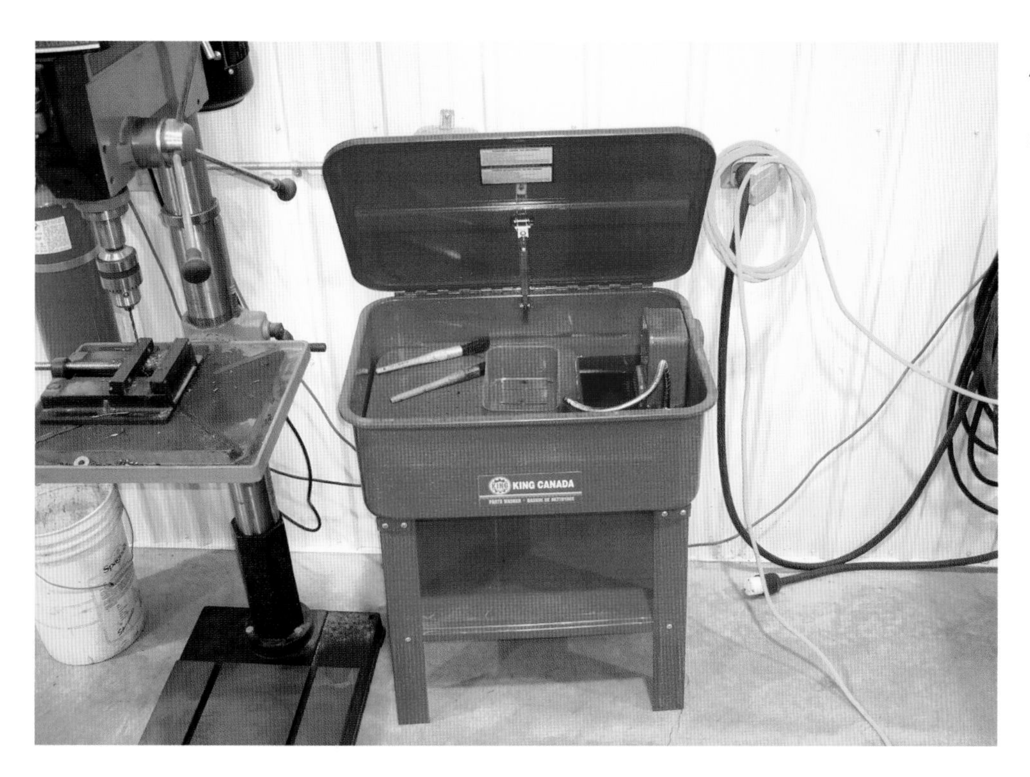

A large floor model solvent parts washer expands the range of parts that can be cleaned.

Although it was common practice in the old days, gasoline should never be used for parts washing because the vapors are too easily ignited. Even the ringing of your cell phone may be enough to spark a gasoline fire if there are a lot of fumes around. If gasoline in an open pan is vigorously sloshed around or sprayed through a compressed air–powered cleaning gun to clean parts, you will create what is essentially a fuel-air explosive. These hyperbaric bombs have been used with devastating effect in recent wars—don't turn your farm workshop into the next demonstration site.

Diesel fuel is sometimes specified for washing diesel engine parts, such as fuel filter bowls, but diesel fuel is much less aromatic than gasoline and takes more heat to ignite.

In the last several years many good aqueous (water-based) alternatives to the usual petroleum-derived volatile organic compound (VOC) solvents have come to the market. VOCs are chemicals that evaporate easily at room temperature. The term "organic" indicates that the compounds contain carbon. VOC exposures are often associated with an odor, but sometime there is no odor. Everyday examples of VOCs include gasoline, low flash-point solvent, nail polish remover (acetone), chloroform, rubbing alcohol, butane, or propane.

Water-based solvents are usually less hazardous to the user than their petroleum-based counterparts. They clean by using a surfactant (such as soap or detergent), a corrosive or alkaline ingredient, or another type of chemical to remove soil from parts. Unlike those of 10 to 15 years ago, most aqueous solutions today are pH neutral so they are much less caustic to skin and surface finishes.

This before-and-after shot shows the power of modern non-VOC rust removing and degreasing products. The cleaner used in this job is not only effective, but also environmentally safe and biodegradable. Courtesy www.rusteco.com

Although aqueous cleaners cost more by volume than petroleum-based cleaners, they generally last 25 to 50 percent longer. Parts-washing equipment with two-stage cleaning, recirculating filtration skimmer to remove oils and other floating contaminants extends solvent life. Case studies document the significant amounts of money saved

Parts cleaning also involves personal cleanup afterwards, so make sure you have hand cleaner and towels nearby.

by heavy equipment and transport fleet workshops users of aqueous cleaners. Proportionally similar savings could be expected in a farm workshop. For tips on how to make aqueous parts washing work in your shop, contact a cleaning equipment supplier or download the fact sheet "Aqueous Parts Cleaning: Best Environmental Practices for Fleet Maintenance" and others from www.dtsc.ca.gov/Pollution-Prevention/VSR/VSR_Fact_Sheets.cfm.

Ultrasonic Cleaning

In some cases, neither type of liquid cleaning solvent will really do the desired job on their own. Some components such as carburetors and injectors have many small passages or orifices that are subject to becoming clogged with gummy deposits from fuel and other contaminants. These parts can be very hard to clean with liquids alone and may be damaged if poked at with stiff wires.

Ultrasonic cleaning is a very useful solution for cleaning such parts. When parts are submerged in the ultrasonic cleaning tank, they are continuously bombarded by ultrasonic energy. When the expansion cycle of the ultrasonic sound wave has enough energy to overcome the surface tension of the liquid, a microscopic cavity is created in the fluid. This cavity grows during the expansion cycle of the wave and collapses during the wave's compression cycle that follows. Rapid formation and compression of the cavities (i.e., cavitation) produces millions of microscopic liquid jets that travel at speeds exceeding 250 miles per hour.

A workshop-size ultrasonic cleaner. The small gear-belt visible at the top of the tank removes any floating oils from the surface of the cleaning fluid. Courtesy Zenith Mfg. & Chemical Corp.

These high-speed jets of liquid gently scrub the parts clean in the smallest recessed part areas and blind holes. An additional result of the cavity's collapse is that the liquid can reach extremely hot temperatures, which also helps free contaminants from the part. By making aqueous cleaners more effective, ultrasonic cleaning also eliminates the need for strong solvents.

Ultrasonic cleaning does have limitations, so you are well advised to choose a unit from a source that can provide ongoing advice on cleaning solutions, power, and frequency for various applications. The parts to be cleaned must be able to fit in the cleaning tank and be covered by cleaning solution. Ultrasonic frequency, power level, and type of cleaning solution that works for steel may damage aluminum parts, especially highly polished ones. Brass, bronze, and copper may require acidic cleaning fluids to remove oxidation and brighten the material. Thick oils and greases absorb ultrasonic energy and may not be removed so the part may need precleaning. Parts with blind holes require proper positioning to ensure they do not trap air. Softer materials, such as plastics, may dampen ultrasonic energy and prevent it from reaching all areas of the part being cleaned.

CHAPTER 10
PRECISION MEASUREMENT

The farm workshop should be equipped with certain specialized measuring tools because, as part of some repair jobs, it is useful to measure if certain parts are still within their service limits. For example, you may be replacing the seals in a leaky hydraulic pump, which appears to be a simple "out with the old and in with the new" task. But if you consult the service manual, you may find that as part of this job you are also advised to check the run-out (deflection from straight) of the shaft. If it's beyond the few thousandths of an inch originally specified, the shaft may wobble so much in use that it will quickly ruin the new seals, so all your labor and the cost of the new seals have gone to waste.

Many other types of jobs call for precision measurement. The pistons in a tractor engine may have a minimum diameter as the wear limit, or the power supply to a computerized fuel injection unit may specify a certain narrow voltage range for proper operation. When your shop is equipped with the right precision measuring tools, you can get away from uncertainty, poor running, more frequent breakdowns, and extra parts costs of the "looks okay to me" approach.

Recent developments have made the cost of acquiring precision measuring equipment a lot more affordable. There is the increased availability of tools manufactured in China and imported for sale at very reasonable prices. But even more than that, there are many used or end-of-line, high-quality tools available because high-throughput, computer-controlled machining and manufacturing is rapidly changing over to non-mechanical tools such as laser micrometers and ultrasonic bolt tension meters. The impact of these two developments in precision measurement tools

Along with newly affordable prices for mechanical precision measuring tools, such as the digital multimeter (left), older technology like the analog multimeter (right) still has benefits. Watch garage sales and auctions for obsolete tool buying opportunites.

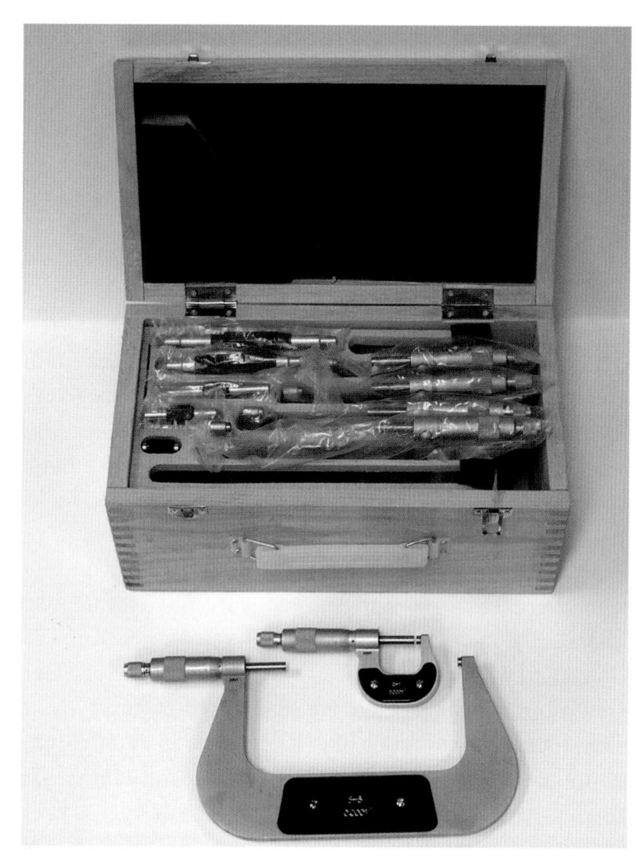

With newly affordable prices for mechanical precision measuring tools, this set of micrometers, including calibration bars for each size, is less than $150.

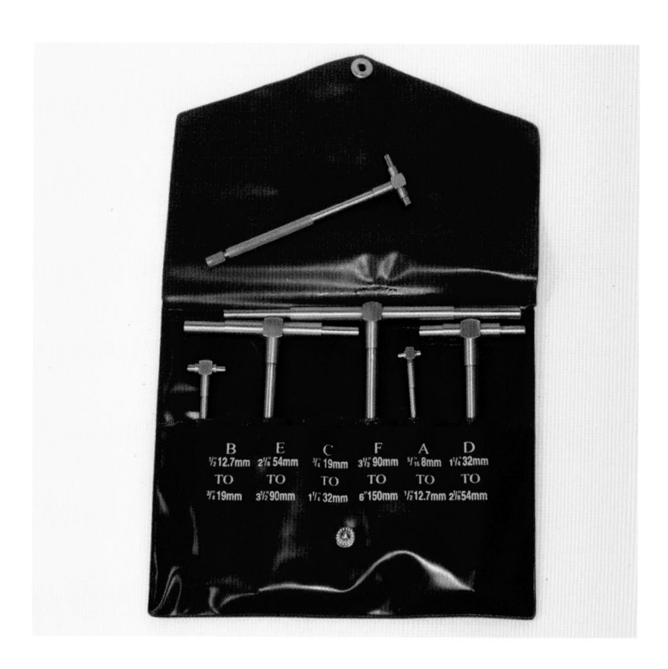

Telescoping gauges can be adjusted to fit internal bores, then read with an external micrometer.

means there is little barrier to any farm workshop being equipped with tools that not that long ago were only to be found in industrial tool rooms.

MICROMETERS

A micrometer measures the size of an object placed between its fixed end (anvil) and moving end (spindle). The spindle moves in and out very slowly thanks to the very fine pitch of the spindle screw: 40 threads per inch for a typical inch-system micrometer or two threads per millimeter for a typical metric micrometer. That means one complete turn of the adjusting thimble moves the inch-system spindle 0.025 inch or 0.5 millimeter, respectively. Fine gradations around the rim of the barrel indicate what fraction of a turn has been used, so the micrometer can measure to three or four decimal places.

Normally a very fine-pitch screw like this would allow hand tightening to exert so much force that it would damage the instrument, the part being measured, or both. Since "hand tight" varies from person to person, it would also introduce unacceptable amounts of error in the measurement. To solve those problems, the end of the thimble will have some system of torque-limiting ratchet. Typically the micrometer's instructions call for turning the spindle until it loosely contacts the barrel, then using the ratchet handle to finish tightening until a certain number of clicks are heard. This indicates the spindle has been tightened to exactly the right amount.

If you acquire a good, used micrometer that no longer has the instructions with it, there are decades worth of machine shop practice books explaining the process and giving many examples. There is also plenty of information available on the Internet—search for "how to read micrometer." These same sources will also have essential tips on useage, such as wiping the measuring tips before use; repeating and averaging measurements to improve accuracy; temperature corrections; and periodic calibration.

One tip common to all types of micrometers is to treat them with care so that corrosion and rough handling do not make them essentially useless. Clean them with a soft cloth after use, store them in their own protective case, keep them away from moisture, and do not allow them to knock around in the toolbox against other tools.

If you're using inch-system measurements, you'll find that precision measurement tools measure in decimal portions of an inch, such as 1.257 inches, rather than fractions as often used in other areas of inch measurement, such as pipe diameter. Mechanic's service literature will often express acceptable limits of part wear or fit in decimals of an inch so that may not be a problem. Sometimes you do need to relate that to a fraction of an inch to determine what size of pipe or wooden block may fit through a hole.

The vernier caliper is a very old design, but it is still effective for general shop measurement that requires more precision than a ruler can provide.

After taking a coarse reading from the main scale, an observation of best alignment on the smaller scale provides accuracy to an additional decimal place.

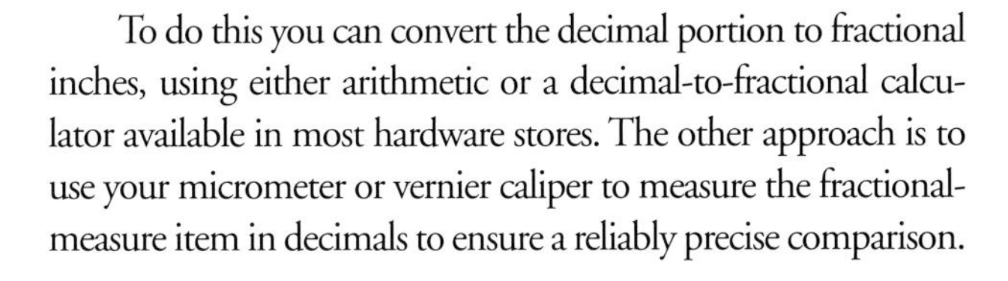

To do this you can convert the decimal portion to fractional inches, using either arithmetic or a decimal-to-fractional calculator available in most hardware stores. The other approach is to use your micrometer or vernier caliper to measure the fractional-measure item in decimals to ensure a reliably precise comparison.

VERNIER CALIPERS

For measurements where accuracy needs are greater than what you can achieve with a tape measure or ruler, but not so critical as to require a micrometer, vernier calipers are an ideal tool. A vernier is a sliding scale placed beside a main scale in order to achieve a greater accuracy than the main scale (e.g., ruler) itself could produce.

The vernier caliper is less precise than a micrometer, but it allows more versatility because it typically can measure three ways with the same pocket-size tool: outside diameters or lengths between the main jaws, inside diameters of a hole with the upper jaws, or depth with the round rod that extends past the main ruler.

One accuracy-limiting feature the vernier caliper generally lacks is the micrometer's ratchet stop to limit the clamping force exerted on the object being measured. A small thumb wheel is provided so you can tighten the caliper only enough to securely contact the item being measured. Over-tightening can bend the caliper or scratch the item being measured.

The vernier reading may be either a straight scale, a dial reading, or a digital readout. Scale-type verniers are more difficult to interpret but are quite inexpensive and robust. Dial-type calipers are much more accurate but more expensive and somewhat more delicate. Calipers with electronic digital readouts are the easiest to use but may lose accuracy as the battery wears down so you'll need to be more careful about determining zero error by calibrating against a known standard, such as the standard length bars often supplied with micrometers.

As with a micrometer, a good book on machine shop practice or an Internet search for "how to use vernier

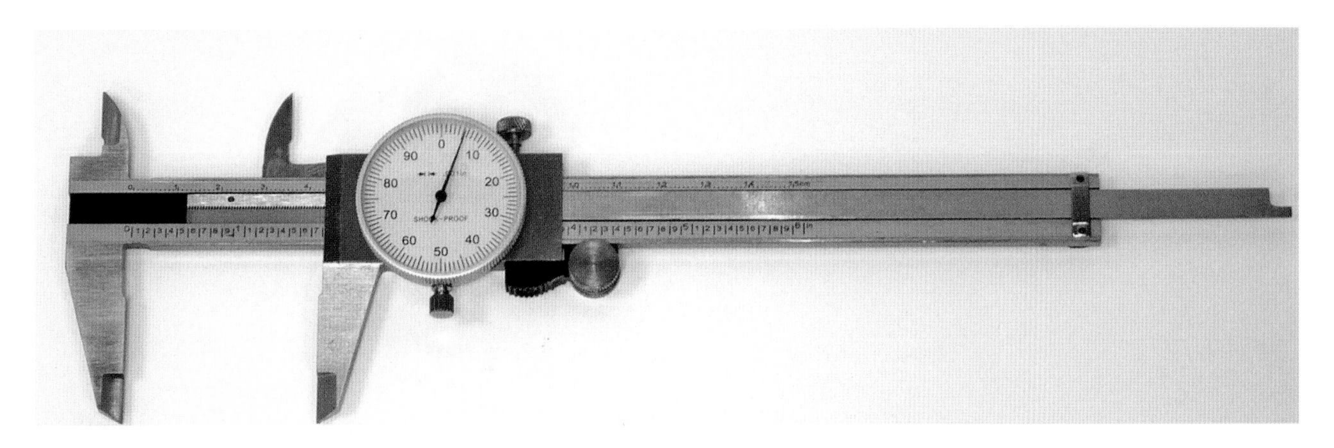

Dial-type calipers make it easier to consistently determine finer measurements.

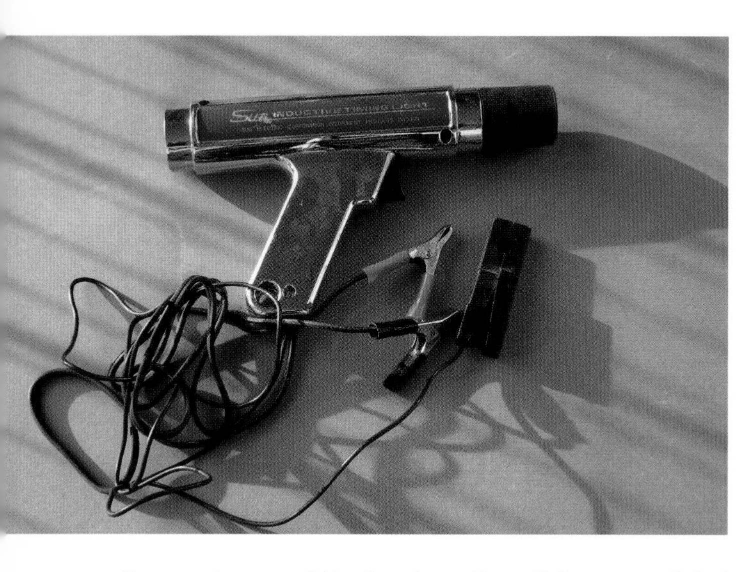

Tune-up meters are useful for diagnosing problems with the non-computerized and mechanical ignition systems in older engines.

The timing light for engines with spark timing can be adjusted or have spark advance/retard mechanisms for different running speeds.

caliper" will give many examples and explain the process for reading them correctly. These same sources will also have essential tips on usage, such as always wiping the caliper jaw faces to remove any dust, grit, or slivers of metal; and storage in a clean, moisture-proof, and shock-resistant case away from other tools.

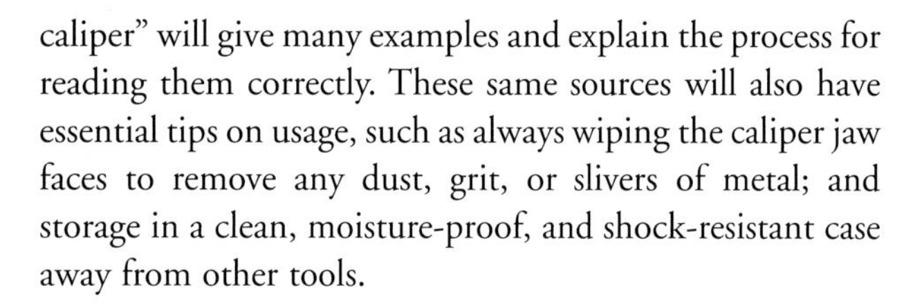

DIAL INDICATORS

Dial indicators display to high precision the distance that a small plunger moves relative to the body of the indicator. In use, the dial indicator is fixed to a stand and the plunger tip is placed against the part to be measured. As the part is rotated or slid back and forth, the dial indicator measures the amount of movement.

Dial indicators of any type can be used for measurements such as end-play (back-and-forth movement) of a shaft mounted in its bearings or run-out (deflection from straight) of a shaft as it turns. For example, using the dial indicator to measure for adequate end play in the crankshaft in a rebuilt tractor engine goes a long way to ensure longer life for the expensive and difficult-to-service main thrust bearings. Using the dial indicator to precisely measure run-out of shafts gives peace of mind that the shaft will turn smoothly and avoid wobbling that leads to vibration that breaks parts, throws belts off their pulleys, or leads to erratic wear of parts that are in contact with the shaft.

The plunger of the dial indicator may move straight in and out or up and down in some types. Traditionally, the plunger is precisely geared to rotate the measuring needle on a dial face, but newer types provide a digital readout. Dial test indicators are similar to dial indicators, but instead of a plunger that moves in and out, they use a small lever arm that moves up and down. These instruments generally have a smaller range of movement and are used for more precise measurements.

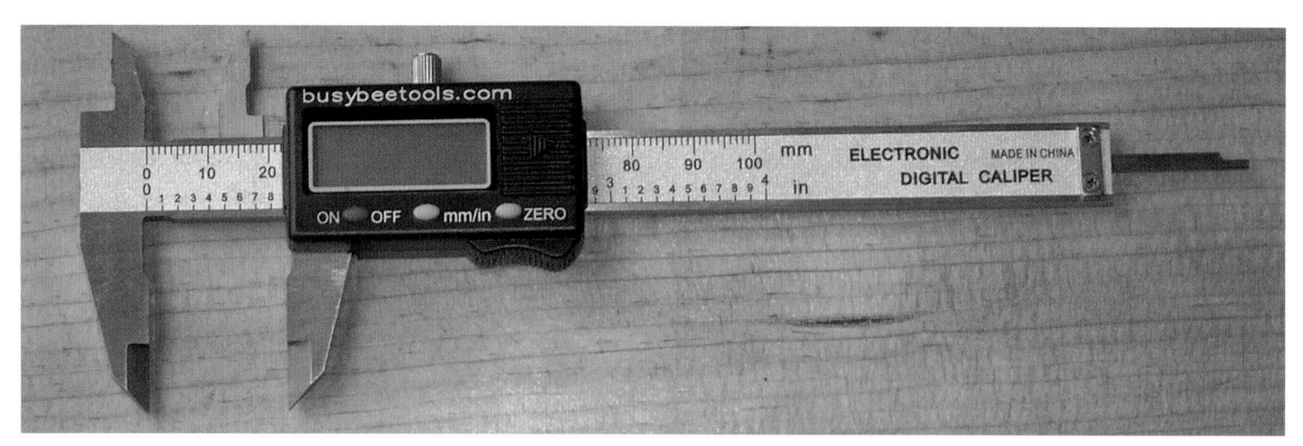

Digital calipers make for a direct readout in either inch or metric and provide a setting to zero, the setting when the jaws are fully closed (or anywhere else you desire).

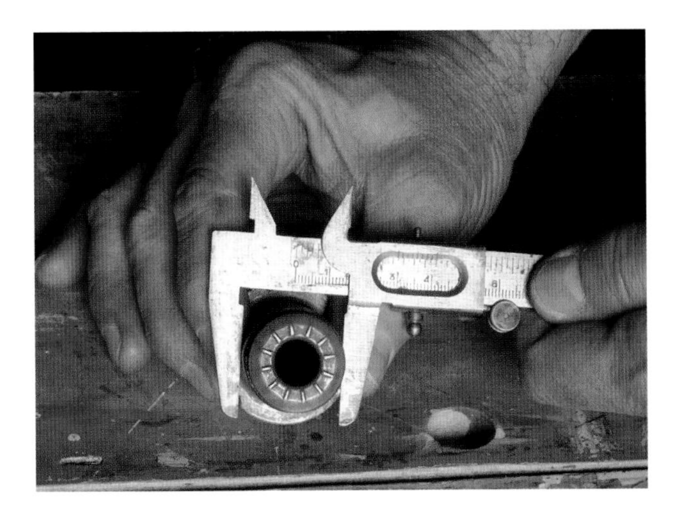

Measure between the jaws for external diameter.

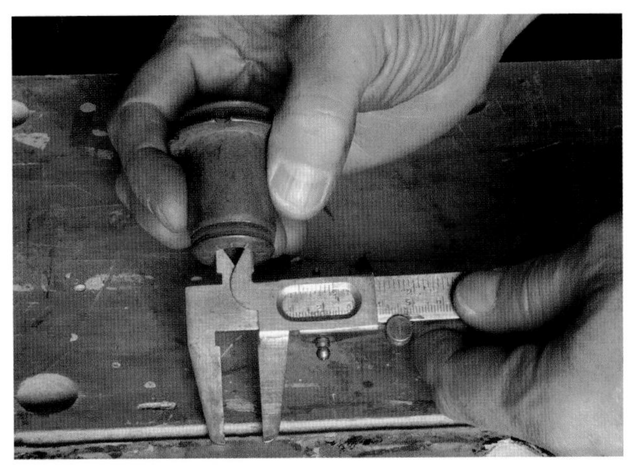

Measure the internal diameter with the upper reversed jaws.

There are many different shapes of points that can be attached to the end of the dial indicator, and each is appropriate to various jobs. A flat point or foot is used to measure convex surfaces, a rounded point is used to measure concave surfaces such as tubing, and roller points may be used for measuring surfaces that move. Straight or curved needle points can reach into holes, while diamond, tungsten carbide, or hard chrome indicator points can be used when the tip is in contact with a material that might wear or corrode it.

Use of the dial indicator depends on it being attached to a solidly fixed stand, so a good stand needs to be part of the dial indicator's equipment. Stands can either mechanically or magnetically clamp to a convenient place near where the measurement is to be taken. For magnetic types, the base has a lever that engages or disengages the magnetic clamping force. These types of bases will only clamp onto steel, cast iron, or other ferrous metals.

The arm of the dial indicator stand provides a way to attach the dial indicator, either with machined dovetails or a mounting lug. Make sure that the stand you choose is compatible with the dial indicator you intend to use. Getting the indicator in the exact position you want it is often difficult, so look for a stand that combines maximum flexibility, ease of adjustment, and a firm hold once the the dial indicator is in the desired position. If the arm jiggles during use, the dial indicator measurement will be unreliable.

For exact details on use, refer to the instructions that come with the instrument or any good book on machine shop practice. An Internet search for "how to use dial indicator" will give many examples and explain the process for reading them correctly. These same sources will also have essential tips on useage.

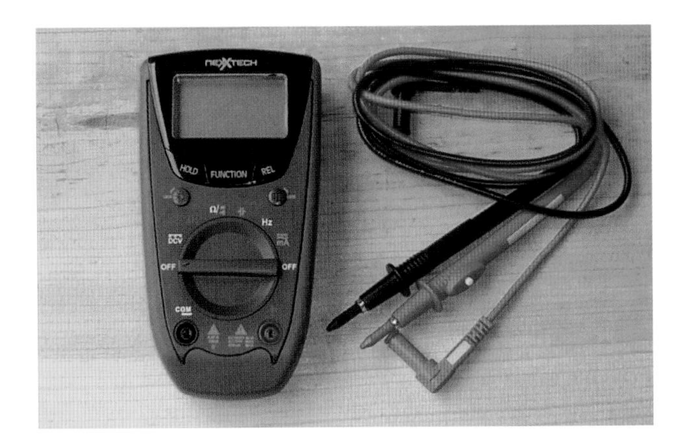

This model of an autoranging multimeter has a setting to audibly announce the reading, which is nice for when you're deep in an instrument panel and don't want to put on your reading glasses to see the reading.

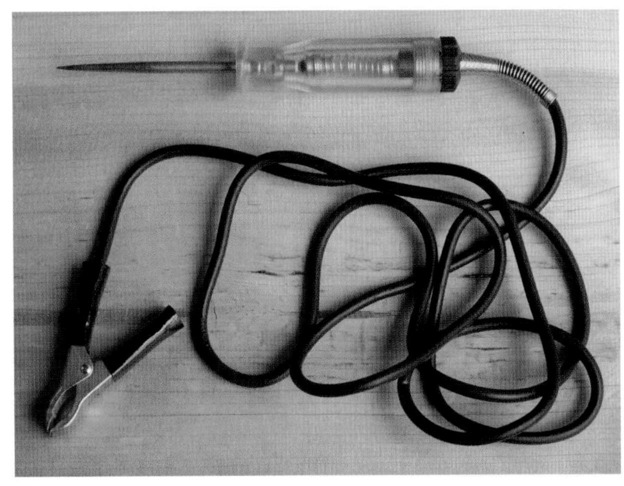

The test light is a truly digital tool—only on or off readings are possible. But if everything you need to trace is the same voltage, it provides a simple method that works well.

Battery hydrometers look like something out of the Model T era (and they are), but they still provide some information on state-of-charge that a multimeter cannot.

TORQUE WRENCHES

Torque wrenches provide a way to measure the amount of twisting force (torque) being applied to a bolt being tightened. Metal bolts are actually slightly elastic and stretch a measurable amount when tightened. That's why bolts in some critical applications (e.g., cylinder heads, wheel bolts) often have a specified tightening force. Too much or too little tightening may cause some sort of expensive and/or dangerous component failure.

Generally any torque wrench scale will be less accurate at the upper and lower reaches of its scale. That's why a well-equipped shop will have torque wrenches in a range of sizes to allow selection of one where the desired torque value falls in the middle two-thirds of the scale. For example, if a maintenance task requires torque in inch-pounds, don't use the low end of a medium-size foot-pounds wrench; get a proper drive unit that reads in inch-pounds. Similarly, if a fastener calls for a high torque value, get a drive torque wrench that has that value in the middle two-thirds of its scale. Do not use extensions to boost the upper end values on a smaller torque wrench.

Refer to the service literature for specific instructions on bolt preparation before applying the specified torque. Specifications may call for dry and clean threads, oiled threads, or the use of anti-seize or sealer. If bolt preparation is not specified, a rule of thumb is that torque values apply to clean bolts with threads lubricated with oil or locking compound. If anti-seize compound is used, the torque value should be reduced by about 20 percent.

No matter what preparation is used, the threads on the bolt should be free of any corrosion or dirt, which will interfere with achieving proper torque. Clean the bolt with a stiff wire brush or by running a die of the appropriate size over the bolt. If the threads that receive the bolt are corroded or dirty, they should also be cleaned by running an appropriately sized tap through first. For nuts or threaded holes that have a hole that goes right through, you can use the tapered taps that normally come with tap-and-die sets. For blind holes that don't go right through the piece, you'll need bottoming taps, which are not tapered and therefore engage threads all the way to the bottom of the hole.

As part of the torque wrench equipment in your shop, provide a place for clean, moisture-proof storage of these precision instruments. Tossing the torque wrench in a tool box with other tools will damage it every time. Post reminders to yourself on how to store the wrench when not in use and the date of its last calibration. Click-type (micrometer) torque wrenches are generally set to zero before storage in order to take pressure off the internal calibrated spring. If pressure is left on, the calibrated spring may weaken to the point of affecting accuracy.

Sometimes the awkward location of a bolt means only an open- or box-end can be used to tighten or loosen it, and there's no room to fit a socket normally used with a torque wrench. For these situations it's possible to buy various adapters that allow the torque wrench to be used with a special stubby open- or box-end wrench. Check at your tool supplier for these adapters and instructions on how to adjust torque readings when using them.

ELECTRICAL MEASUREMENT TOOLS

Tools for precise measurement of electrical system conditions are part of a well-equipped shop. A bit of time spent with a multimeter or battery tester can save hours or days of repair time on a tractor that's running poorly or an electrically monitored implement that will not produce satisfactory results.

Many people who are otherwise enthusiastic about working on machines are extremely reluctant to tackle anything involving electrical wiring or components and cite these problems as particularly illogical or baffling. The right electrical measurement tools can go a long way to dispersing these concerns. Electricity does inevitably conform to simple well-known physical laws, even more so than most mechanical systems. Skill at diagnosing and repairing electrical faults can go a long way in keeping your equipment running. Many find the ability to solve "electrickery" problems a rare and impressive bit of skill that approaches the level of wizardry! It is best to buy those electrical measurement tools and learn how to use them.

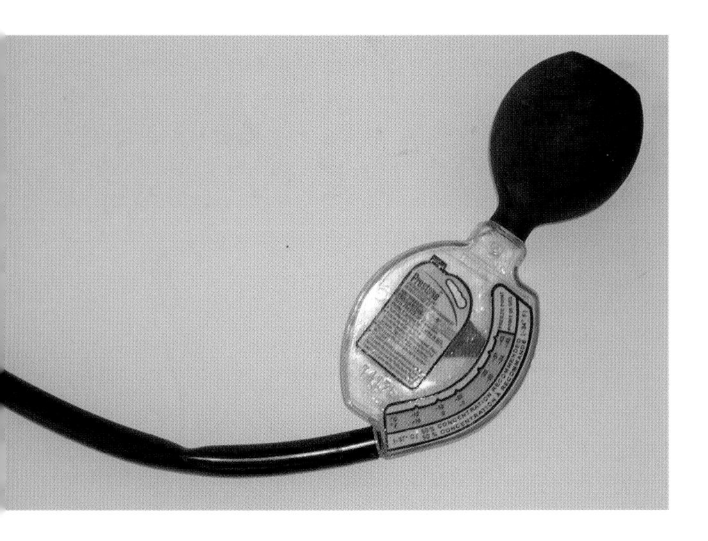

The coolant hydrometer works on the same principle, but should only be used for measuring coolant concentration. The readings of the two different kinds of hydrometers will be of no use when not used for their intended systems.

A multimeter can easily measure power supplied (volts), resistance to flow of current (ohms), and current (amps). The advantage of equipping your shop with a multimeter rather than a simple test light for measurement is that the multimeter can tell you exact voltage drops, resistance changes, or current flows at various points in the wire, or through an electrical component. The test light only gives a "yes or no" indication of voltage.

Even without a wiring diagram or detailed electrical specifications, you can use the multimeter to narrow down the area that's creating a problem. For example, if a tractor's starter or any of the instruments fail to operate properly, it's fairly simple to use the voltage (V) measurement function to determine if there is any difference in voltage at one end of a wire and the other. If there is, checking voltage at various points along the wire, such as between the wire and connector or connector and ground, can quickly pinpoint the fault. An alternative or as a confirmation method, you can use the resistance setting to check that a wire is able to carry current from one end to the other without resistance. Resistance is measured in ohms, the symbol shaped like a horseshoe. For more practical examples on using the multimeter to find and isolate faults, refer to the book *How to Keep Your Tractor Running.*

Choosing a multimeter for the workshop comes down to choice between readout type and range selection. Digital-readout multimeters are nowadays more widely available than older analog (dial type) and are more precise in most cases. But if you prefer working with analog instruments and can find a good-quality used analog unit in good working order, it can give satisfactory results.

The load tester gives a quick indication of battery life, but don't leave it connected for more than a few seconds because of the enormous heat buildup in the tester during use.

You may also find an analog meter particularly useful if you want to measure very low amperages, such as when trying to track down what is draining a vehicle battery when all switches are off. Sometimes a short circuit may still draw a few thousandths of an amp and the easy-to-see "twitch" of an analog meter needle will help you detect it where a digital meter will not.

In terms of setting the measurement range on the multimeters, ones with manually set ranges are more common and less expensive, but the auto-ranging type is rapidly becoming an inexpensive and easier-to-use alternative. With a manually set multimeter, you need to set the desired function (volts) and select the range of measurement based on the maximum expected reading (e.g., 20 volts or 20 millivolts).

With an auto-ranging multimeter, you set the desired function, and the multimeter automatically chooses the correct range so there is no need for additional interpretation of the displayed value. On manual-ranging meters (analog or digital), you set the range dial to a value that will include the highest expected reading. Many find the proliferation of ranges on the multimeter's face confusing, so an auto-ranging type may be preferable.

Other electrical measurement tools include those needed for battery testing. The battery hydrometer measures the specific gravity of the electrolyte in batteries, which indicates how much charge remains in a battery. Note that battery hydrometers and coolant hydrometers are

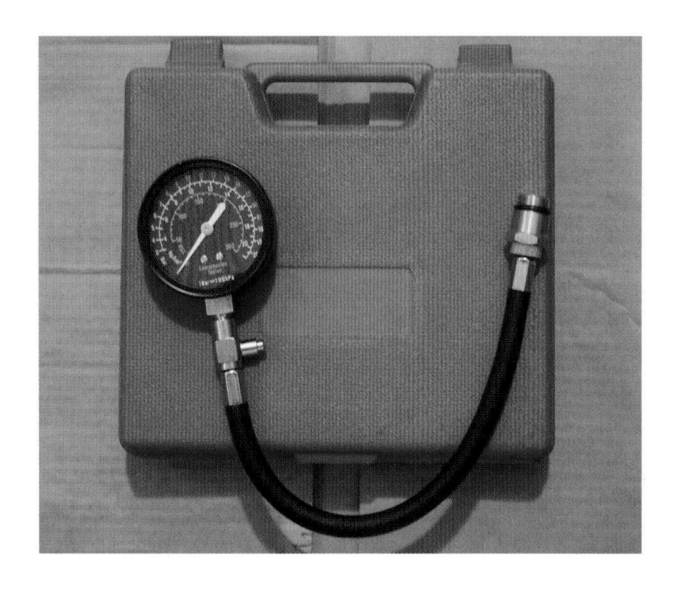

To replace reliance on a finger over the spark plug hole to assess compression, an inexpensive compression test gauge gives accurate results.

completely different instruments and one cannot successfully be used in place of the other.

The battery load tester verifies that the battery can reliably deliver its specified power. Using the load tester is always useful in determining whether it's worth recharging a battery or whether it's time for a replacement. The tester is especially useful as equipment in a farm shop because it eliminates having to remove a battery (or set of batteries in heavy-duty equipment) and make a trip to have the load test done at a battery shop.

For working on older gasoline engines that have point-and-condenser ignitions and variable timing—such as those found in older farm tractors—an ignition diagnostic meter and strobe-type timing light will be necessary parts of your shop equipment. These multiple functions of the diagnostic meter are key to making several adjustments that will keep the engine starting and running smoothly. In particular, the dwell angle function of the diagnostic meter allows precise indication of how long the ignition points stay open. The tachometer function makes precise tuning of carburetor idle and low-speed and high-speed fuel mixture screws a lot easier. The timing light permits a precise setting of when the spark plugs fire and are used in conjunction with the diagnostic meter's tachometer. It also permits a check of whether the distributor's spark advance mechanism is functioning correctly. The multi-function diagnostic meters and strobe timing lights made for automotive use are also capable of handling tune-up task for gas-engined tractors and other farm equipment.

Like other precision measurement tools, precision electrical measurement tools should be stored in a clean, dry place separate from other tools in order to avoid damage. With multimeters, the internal battery should be removed when not used for long periods of time because aging batteries can release corrosion that fouls the instrument. The central glass tube of the hydrometer is vulnerable to breakage due to rough handling. The hydrometer should also be flushed with clean distilled water and hung up to dry after use so that electrolyte doesn't drip out and cause corrosion.

Tune-up meters are useful for diagnosing problems with the non-computerized and mechanical ignition systems in older engines.

The timing light for engines with spark timing can be adjusted or have spark advance/retard mechanisms for different running speeds.

CHAPTER 11
METALWORKING AREA EQUIPMENT

Today's farm workshop can easily, and at moderate cost, include metalworking and fabricating equipment capable of producing and repairing high-precision products. Typical projects could include anything from building a new trailer to fabricating a mount for a hydraulic motor. Machines in this area cut, bend, drill, mill, and grind metal to required shapes in preparation for use or for welding and bolting into further structures.

LAYOUT AREA

Achieving precise results when working with metal has to start with marking out the rough stock as precisely as possible. The layout area you need depends on the size of the project, but essential components include a flat, stable surface (floor or bench top) and something to hold the piece steady while you mark it. It's quite frustrating to have steel tubes roll away while both hands are occupied holding a tape measure and a marker.

If you're marking a small piece at a bench top, a vise or clamps may be all you need. If you're working with larger pieces on the floor, heavy sandbags and wooden blocks help keep things still while you measure and mark.

Before drilling metal it's usually necessary to make a punch mark on the spot to be drilled. This centers the drill point in the hole and helps make sure it does not skitter off the mark. With an automatic center punch, you press the point down on the mark and at a preset pressure, the internal spring trips and drives a striker against the tip that punches a mark into the metal. If you're working with hard metals and want to slightly enlarge the locating hole before drilling, you can strike on the mark with a manual center punch and heavy hammer.

HAMMERS, FILES, AND PULLERS

Your workshop should be equipped with several sizes of metalworking hammers, which differ from claw hammers—as used in carpentry—quite a bit in shape, weight, and striking face hardness. Carpentry hammers are made for striking nails made out of metal softer than the hammer's head, and metalworking hammers are made to handle harder metals. Carpentry hammers should not be used for metalworking unless you have nothing else on hand.

The ball-peen hammer (also known as ball-pein, engineers, or machinists hammer) has a flat striking face on one end and a round striking face on the other. The round face imparts a rounded shape to the metal being formed, while the flat face is useful for all types of metal striking work. The original function of the round face was to peen (mushroom) protruding metal, such as rivets on cutter-bar sections or the ends of roller chain connector links. Hammers are graded by the weight of the head. They can be bought individually as needed, or in sets including 8-, 12-, 16-, 24-, and 32-ounce sizes, which should take care of the needs in any farm workshop.

Other useful types of metalworking hammers include the cross peen, clubbing, dead blow, soft face, and slide

Pencils don't work that well on metal, but after spraying with layout dye, a light score mark with a carbide-tipped scriber shows up as distinctly as a line on a blueprint.

An automatic center punch frees the user from needing both hands to hold a hammer and manual center punch.

types. The cross-peen hammer has one flat face and one face that tapers perpendicular to the handle. The pointed face is still useful for hammering straight-line bends in metal, expanding the ends of rivets in an X-pattern, or driving metal pieces sandwiched between other flat pieces.

The clubbing hammer has a heavy head and short handle, and the flat faces are larger than those of the ball-peen or cross-peen types. The larger size of the flat faces makes it easier to strike punches and chisels without missing, while the heavy head and short handle means you can impart lots of striking force using a slower, shorter, easier-to-control swing. Taking a mighty swing with other types of hammers missing the punch, and smacking your wrist or fingers is a very painful experience. The clubbing hammer looks simple, but is carefully designed to avoid that kind of problem.

The dead blow hammer is designed so that it does not bounce after an object is struck. This is very handy in case you have to strike a very hard shaft in order to loosen it. The soft-face hammer allows you to apply striking force to metal that might suffer damage from an ordinary hard hammer face. This is very useful when you need to loosen a cylinder head that is stuck to the block. The soft-face hammer often has a choice of brass, rawhide, or plastic faces and takes the place of having to hold a softwood block between the hammer and the part being struck.

The slide hammer can be used for removing parts that are stuck tightly together, such as bearings and the casting that steadies them. One end is hooked to the part, and then

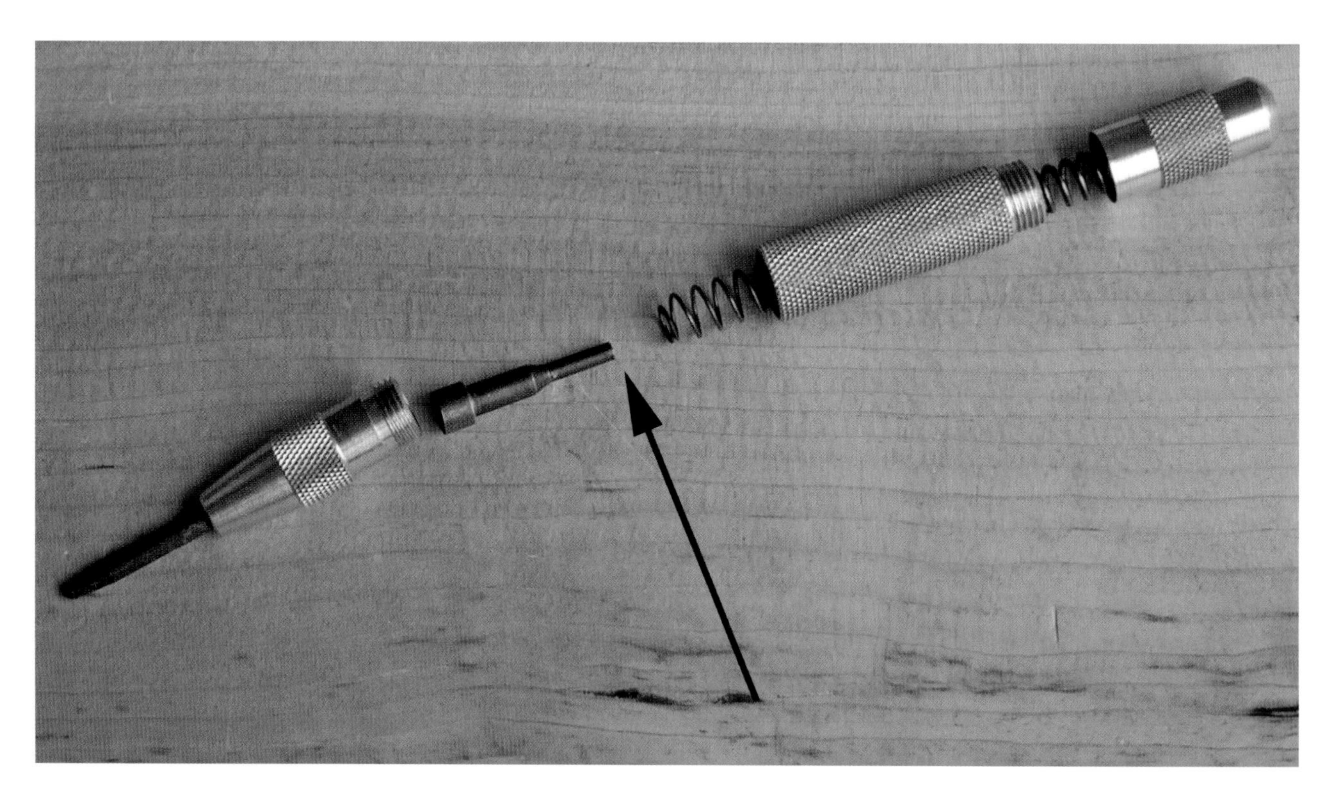

Lower-priced automatic center punches may quit working after being used for a while. The problem is that the trip surface indicated becomes slightly rounded off. Carefully file it flat and smooth again to restore functionality. More expensive types of automatic center punches have better steel and often have an adjustable striking force.

Abrasive cut-off saws are quick and easy to use. They are also more accurate in terms of achieving a square cut than a handheld hacksaw.

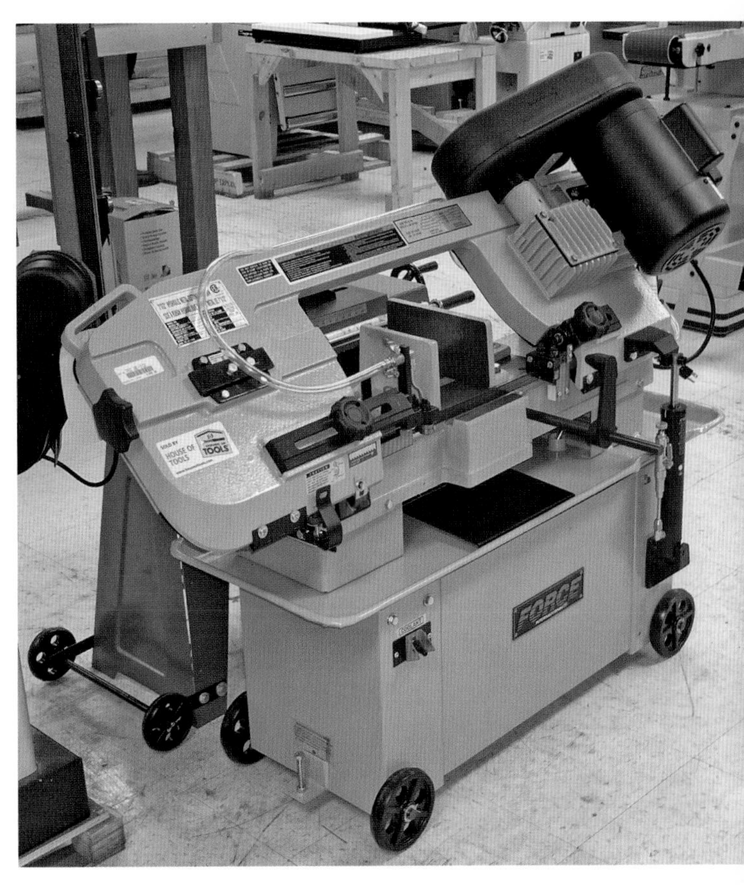

The metal-cutting band saw has become the new standard tool for cutting with some degree of precision. Saws are available in a wide variety of size and configurations to handle any shop need.

the weight is slid rapidly along the shaft until it strikes the stop and jerks the whole assembly outwards. This tool is tricky to use correctly. Although service manuals sometimes specify use of a slide hammer for particular procedures, screw-type pullers are often used instead to achieve a more controlled pull with less shock to the part.

If you do any work on small engines as used in lawn tractors or irrigation pumps, do not use either impact hammers or screw-type pullers for flywheel removal unless the service manual specifically says it's okay. Flywheels are typically removed to replace the soft metal shear key that protects the crankshaft, connecting rod, and piston from damage if the engine is forced to a sudden stop, such as when a lawnmower blade strikes a rock or an irrigation pump suddenly seizes. The key must be replaced to reset the ignition timing correctly otherwise the engine may never start or run properly. On these engines, a specialized flywheel removal tool must be used to reduce risk of breakage of the relatively fragile cast flywheel. These tools fit over the end of the crankshaft stub once the retaining nut has been removed to allow you to tap on the stub while gently prying the flywheel away from the engine. The tool is usually only a few dollars and available at engine parts suppliers and well-stocked hardware stores.

Files are used for shaving, smoothing, and fitting metal parts, and for basic sharpening, such as with axes and lawnmower blades. For farm workshop metalwork, a set of flat, round, and triangular double-cut (grooves running both ways) machine files and single-cut (grooves running only one way) mill files will do the job. Mill files leave a smoother finish than machine files but cut more slowly. Files are made in four levels of coarseness: coarse, bastard, second, and smooth. Again, the finer cuts are slower. File sets often come with a "farmer's friend" or mower file with a handle on the end. These files were originally used for sharpening hay mower blades and you might find that you end up using all of them for any repair job.

A few additional files will be very useful in farm workshops. Needle files are smaller, which make them useful for cleanout work in tight places. The auger bit file is useful for filing right up against an edge you don't want disturbed. A roll of emery cloth (abrasive on a strong cloth tape) allows

Metal work involves applying considerable force to the work piece, so equip your shop with a strong vise to hold the work.

filing and polishing round shafts or holes without disturbing the diameter very much. Strips of tape can also be torn off and glued to solid backing to make any shape of file you need for a unique situation.

THREADING EQUIPMENT

In the course of repairing and maintaining equipment in the farm workshop, threads on bolts or threaded holes are often damaged or bent. A tap and die set provides the means to restore damaged threads with either a die for external threads (bolts, studs, screws) or a tap for internal threads (nuts or threaded holes). Taps and dies can also be used to cut new threads into or onto metal where required. Be sure to go easy on the muscle power and use plenty of tapping lubricant when doing so because the hard steel in taps and dies is very brittle and can snap easily if forced. Removing that hard chunk of tool steel from a hole will be very difficult, if not impossible.

Sets of thread restoration tools are also available, and while they will not cut new threads, they do provide more flexibility in restoring threads. The thread restoring files in these sets are extremely useful because they will work on very large-diameter threaded objects that taps and dies are too small to handle. The file is passed gently over the threads and as the teeth engage neighboring undamaged threads, they cut away the damaged parts. Thread restoring files are also available, separately.

Another option for restoring damaged internal threads, such as on a stripped spark plug hole, is to use a thread insert (Heli-coil) kit. These kits provide a drill bit sized slightly larger than the old hole, and coils of diamond-shaped steel wire. The threads of the damaged hole are drilled out and the coil of wire is fixed into the enlarged hole to form new threads. Often this repair is stronger than the original threads, especially in aluminum-alloy castings.

CUTTING METAL TO LENGTH

There are three kinds of machines typically used for cutting large pieces to tube or bar to length: the abrasive-wheel cutoff saw, the power hacksaw, and the band saw. Along with the powered cutting machinery, the basic hand-powered hacksaw is still occasionally useful in the shop and should be stocked with a selection of blades for various metals.

The abrasive cutoff saw is widely used for initial cutting of steel parts. It's fast and inexpensive but also quite noisy and throws off a lot of sparks and slag around the shop. The large burr left on the ends of the cut pieces also needs to be ground or filed off before welding. The large-diameter abrasive discs used in this saw are brittle and can break easily if they twist because the work piece is not securely clamped during cutting. Fortunately, they are not very expensive so breakage is less an issue of cost than is the time needed to change discs.

A power hacksaw provides a more precise cut with less noise than a cutoff saw but it cuts more slowly. Blades can be relatively easily and quickly changed to suit the material being cut. The frame holding the blade should be strongly built so that the blade does not twist, which makes the cut wander off-course and may snap the blade. Controls to regulate cutting speed and feed pressure should be easy to operate and clearly understandable because an improper setting can

Seeing a workshop with a lathe immediately indicates that the owner is capable of some pretty sophisticated fabrication work.

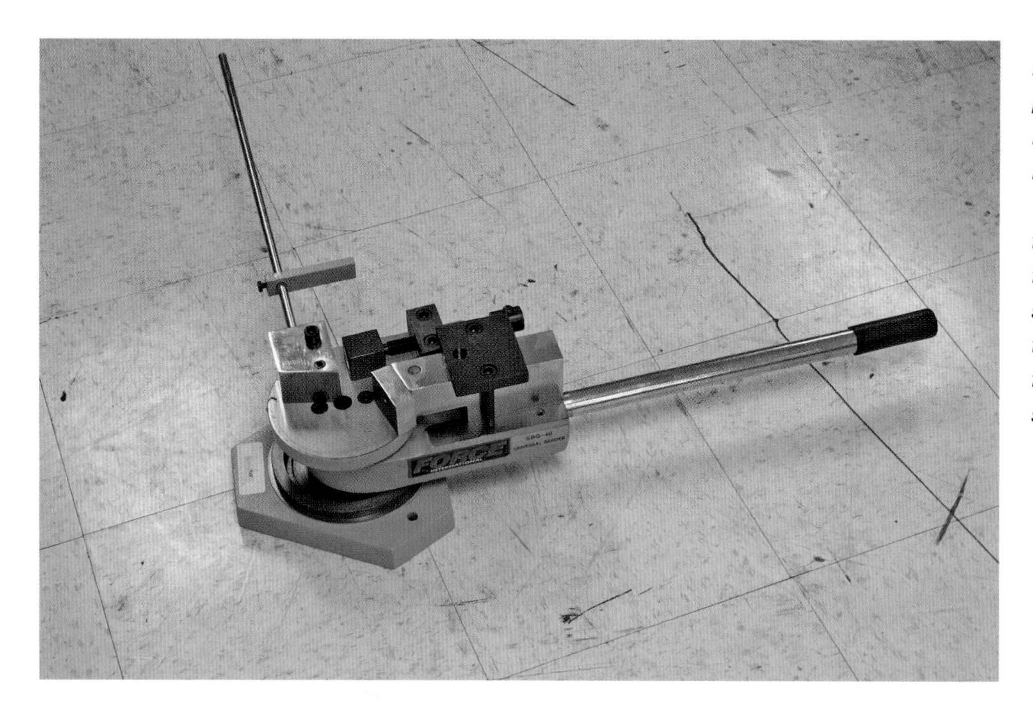

Fabricating items from bent tube is possible on a small scale by hand methods, but becomes easy when leverage is applied with a bender. The benefit of a bender is that it forms the bend without flattening a tube or letting the sides of solid stock bulge out. It also allows you to make a bend without the need for heating, which could harden or soften the metal once cooled.

make the blade jam into the material and possibly break. Provision of a coolant/lubricant feed can help make cutting faster, quieter, and less wearing on blades.

The horizontal, vertical, or combination band saw has now mostly taken over from the power hacksaw as the preferred tool for precise cutting to length. It's easier to start the cut exactly at where you've marked and cutting tends to wander less. Handheld band saws are now available for cutting jobs where you can't fit the work into the frame of floor-mounted models.

A standard horizontal or vertical saw orientation is suitable for round work pieces, while having a bit of forward lean (cant) in the blade orientation provides an advantage when cutting square or rectangular material. A canted blade cuts approximately 30 percent faster, keeps a more consistent path, and is less wearing on the blade.

For farm workshop saws that need to be good at cutting both round and rectangular metal, a forward-canted blade is a good choice. It maximizes efficiency when cutting square or rectangular pieces while imposing very little performance penalty when cutting round stock.

A horizontal band saw is generally less costly than a vertical machine and will provide excellent performance for straight cuts and the occasional miter (angled cut). If you expect to be making a lot of miters, a vertical band saw is the best choice. A vertical saw allows you to position the work piece vises closer to the cut, which improves blade stability and accuracy of cutting. Keeping the vises closer to the cutting zone also dampens vibrations that produce screeching sounds during the cutting process.

When evaluating the coolant/lubricant system on a saw for your shop, keep in mind that flood- or drip-type coolant

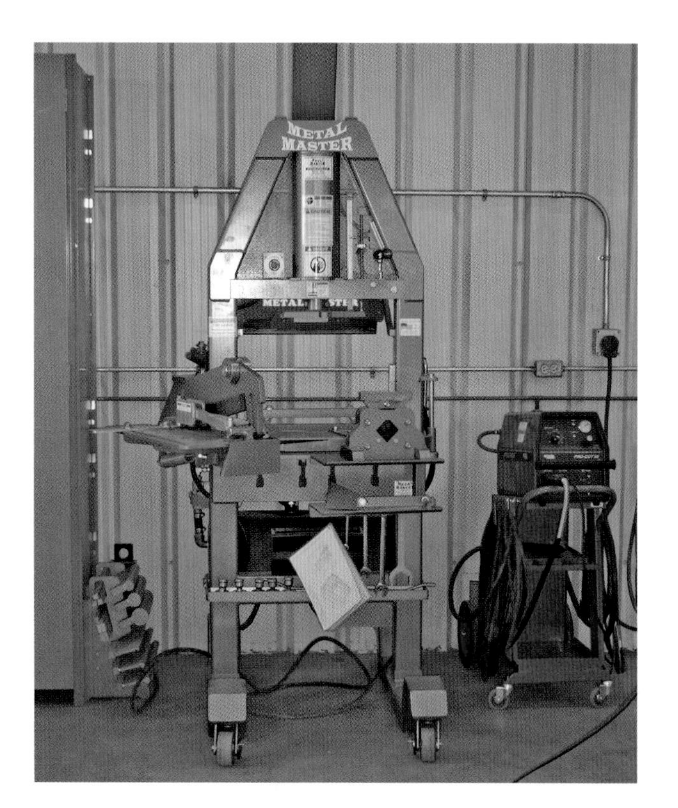

This large metal shaper is essentially a cold forge and replaces the heat and hammering of blacksmithing with enormous amounts of precisely controlled hydraulic power. The tools visible at lower left provide various kinds of shapes to be formed.

With a personalized shop plan developed from Chapter 1, you can make an informed decision on which size of drill press is best for your shop.

Moving the drive belt higher or lower on the stepped pulleys of the drill press controls the drilling speed.

systems provide heat transfer, lubrication, and blade life, better than mist systems. However, coolant does tend to drip onto the band saw frame and spill onto the floor, which makes a mess unless catch pans are placed underneath the machine.

Mist systems are cleaner with less dripping than conventional flood systems and are quite effective when cutting tubing or pipe. However, blade life can be shorter when using a mist system instead of a flood system.

Feed pressure regulation is very desirable on a band saw. It keeps the blade from jamming in the cut and also helps break in new blades to maximize blade life. Points are extremely sharp on the teeth of a brand-new blade. If you start cutting with full feed pressure, tiny chips will break off the cutting edge and prematurely damage the tooth point, similar to the way a freshly sharpened pencil point will snap off. A minute or two of cutting at recommended speed but one-half the recommended force will reduce chipping and extend the blade's life.

Occasionally check the shavings as your saw cuts. Ideally you should be seeing evenly shaped spiraled curls of metal coming off the cut. Tightly curled, brown to black shavings that are warm to the touch are a sign of too much down pressure, while thin chips indicate insufficient feed pressure. If the cutting speed is too high you will get chips that have a blue hue and feel hot to the touch.

If you use a band saw for metal, a blade repair kit is good to have in your workshop. These kits have a small jig to hold the ends of a broken band at just the right gap for brazing it back together into a usable blade.

GRINDERS

Abrasive-wheel grinders, both permanently mounted and portable handheld types, will find plenty of use in shaping metal, beveling edges to prepare for welding, removing burrs, and removing the small annealed zone that results from metal being cut with an oxyacetylene torch. In

A small vise on the table of the drill press keeps the work steady during drilling, especially in the critical last part of the cut.

addition, the wire wheel that is permanently mounted on one end of a stationary grinder (or temporarily mounted on a portable grinder) is very useful for removing paint and corrosion or achieving a satin-type finish on metal.

For a stationary grinder, mounting on a freestanding heavy base or pedestal provides maximum freedom to move work pieces relative to the grinding wheels. Grinders can also be mounted on the bench top but it does restrict movement of the work pieces.

A specialized type of stationary grinder is used to form the curved "fish mouth" needed where one hollow tube intersects another in a metal structure. You can observe these types of mitered metal joints in any modern bicycle frame. To make the fish mouth, a tube is pushed against the spinning vertical spindle of the miter grinder until sufficient curvature is achieved to let the wall of the shaped tube fit snugly against the outside wall of the tube it will touch. This maximizes the area available for welding and therefore results in the strongest possible joint. If you're making only an occasional miter joint, there are moderately priced kits that allow you to accomplish the necessary shaping by using a metal-cutting hole saw in your drill press; the miter grinder does the job a lot more efficiently.

When grinding away excess metal is the goal, a large grinder with plenty of torque and low energy consumption is the most preferable. Ratings may be in terms of wheel size or output power, which is the power available for tool operation after heat loss. Grinders with features such as low-friction ball, roller, or needle bearings, high-efficiency fans with protective screens, and cords with plenty of insulation generally result in less electrical power lost as heat, which means more of the power consumed is turned into useful grinding work.

However, some jobs also call for more finesse and slower rates of grinding. That's when it's also nice to have a small, low-powered grinder. With a handheld portable grinder, the light weight also makes it less fatiguing to use when the job doesn't call for hogging out great chunks of material. Many small, low-priced grinders can now be found for under $40, so it's easy to make these part of your farm workshop equipment.

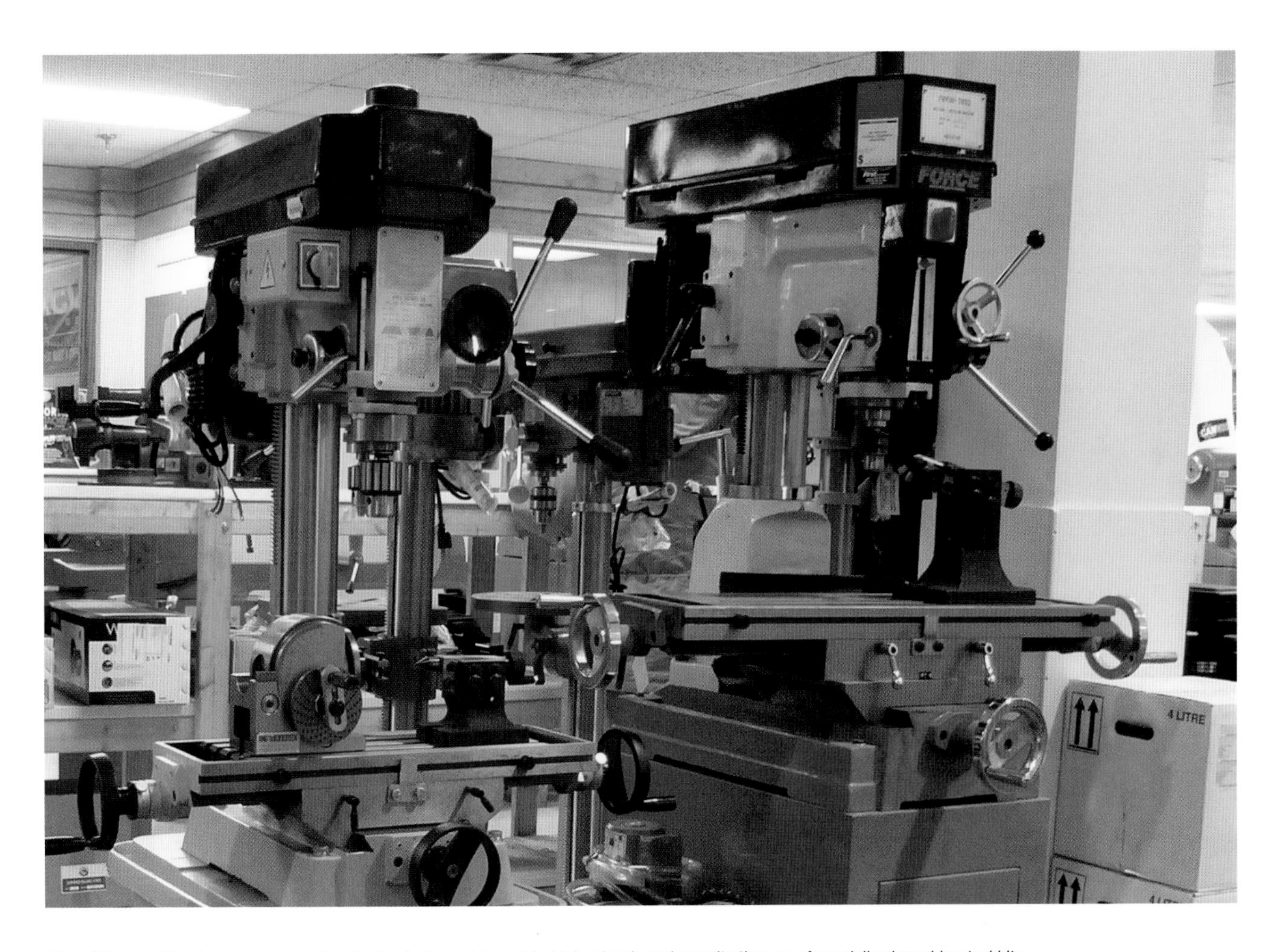

The milling machine takes accuracy of vertical-spindle metal work to higher levels and permits the use of specialized machine-tool bits.

With any type of grinder, look for safety features such as sturdy, easy-to-adjust work piece rests, grinding wheel guards and spark shields. The easier it is to adjust rests, guards, and shields, the more likely you'll use them and gain the benefits of improved safety. Ergonomic design features reduce fatigue and frustration, especially when you have to use a handheld grinder in an awkward position.

On a large grinder, a slip clutch is a feature that protects both the operator and the tool if the wheel makes a sudden stop. A grinder spinning at 10,000 rpm has a tremendous amount of momentum in the abrasive wheel, electric motor, and any gears in between. If the work piece gets stuck between the wheel and a carelessly adjusted work piece rest, or a handheld grinder is pressed too hard against the work, several bad things happen. In a small direct-drive grinder, the armature stops and strains the electrical components. But with a heavy-duty tool, the work piece may be wrenched violently out of your hands and hurled in an unexpected direction, the whole grinder may try to lurch to the floor or kick back out of your hands, the expensive grinding wheel may shatter, and any gears or drive pins between the wheel and grinder motor may be broken. A slip clutch is valuable in preventing this by allowing the whole mechanism to safely adjust to a sudden accidental stop and prevent damage to both the operator and the tool.

The on-off switches on grinders are designed so it takes constant finger pressure to keep the grinder turning. This dead man switch is an important safety feature because if you drop or let go of the grinder, it spins to a stop. Never tape the switch down or override the dead man feature. Overriding the feature may sometimes seem convenient, but in case of an accidental release it could result in the grinder whirling across the floor and damaging you, itself, and other equipment.

Once you've chosen the grinders for your shop, equipping them with the right abrasive wheels is also a good idea. Using the same wheel for everything can result in metal that's not as well finished as it could be, as well as more wear and tear on both the grinder and the operator. Abrasive wheels are generally made from three different materials:

The combination machine puts both horizontal gap-bed lathe and vertical milling/drilling machine in one versatile unit. Various sizes are available from small hobby units to large submarine machinist-size models.

1. Aluminum oxide for ferrous metals (e.g., steel) and alloys with high tensile strength.

2. Zirconia alumina for grinding or severe-duty grinding on ferrous metals and alloys with high tensile strength.

3. Silicon carbide for nonferrous metals such as copper, aluminum, bronze, and ferrous metals with low tensile strength, such as cast iron and ductile iron.

For large grinding jobs, match the hardness of the wheel to the job rather than whatever came with the grinder. Softer and coarser wheels cut faster but wear out quicker and leave a rougher finish, while harder or finer grained wheels last longer but grind more slowly. The letter-number code on the wheel indicates its composition and use. The first letter is the designation of the material: A for aluminum oxide, ZA for zirconia alumina, and C for silicon carbide. Next, a number indicates grit size: the smaller the number, the coarser the grit, which is similar to sandpaper ratings. A letter then indicates hardness: A is the softest and Z the hardest. For handheld grinder wheels, two more letters may be used to indicate the type of resin used for bonding and the reinforcing material used in the wheel: B indicates a phenolic resin and F indicates fiberglass reinforcement.

For all types of grinder wheels, match the wheel size and revolutions per minute, or rpm, rating to your grinder to prevent explosive failure that ejects broken pieces at high speed. For safety's sake, do not run an oversized wheel on a grinder with the guard removed because this severely increases risk to the operator.

DRILLING

The stationary drill press (also known as a pillar or pedestal drill) is a vital part of your workshop's metal working equipment, more so than the usual corded and cordless handheld drills. Compared to a handheld drill, the drill press vastly increases the precision and speed of drilling, which makes it an important part of your shop equipment. Because the moveable drilling unit is held exactly parallel to the center

A rack for metal stock lets you easily store and retrieve the pieces needed for projects.

post, you can be sure that holes are drilled vertically to a depth you can control precisely with the crank handles. The tilting and swiveling table makes it possible to drill at an angle if necessary. A variable speed system makes it possible to match drilling speed to the material being drilled.

Both large freestanding and small bench-top drill presses are very handy to have in the shop. You may find the small units, often referred to as hobby machines, handier to use for small jobs and ordinary metals where high torque drilling is not necessary.

The high-torque drilling that makes drill presses so useful also contributes to one very common safety problem: the work piece being jerked out of the operator's hands if the bit jams in the hole. This is most common when the operator just wants to quickly drill a hole and uses a lot of pressure without clamping down the work piece. The result is often skinned knuckles and cut skin as the unsecured work piece spins around. The solution is to always clamp the work piece when using the drill press for metal; drill with only firm, steady pressure on the bit; and use lubricant in the hole.

Drill press vises clamp or bolt to the table of the drill press to hold the work exactly in place. Some drill press vises simply clamp the work, some allow you also to tilt the work piece to make angled holes, and some have calibrated hand wheels to allow precise positioning of the work piece back-and-forth and side-to-side. None of these vises are very expensive and all of them can help you achieve impressive results with substantially increased safety and convenience.

The drill press can use any of the round-shanked general-purpose bits normally used with handheld drills. In addition, several other types will be useful for particular types of drilling. Cobalt-alloy drill bits hold their hardness against heat in tough drill situations so they are used for drilling hard metals, such as stainless steel and titanium alloys. For drilling holes larger than 1/2 inch in diameter, Silver & Deming–type drill bits (also known as blacksmith or prentice bits) are used. They have a 3-inch-long diameter section for drilling, with another 3-inch-long shank, usually 1/2 diameter, to fit a standard chuck. Titanium bits are steel bits with a titanium nitride (TiN) coating, which reduces friction and keeps the cutting edge sharp, up to six

Keeping your shop's metalworking equipment close together makes for a more efficient operation.

Roller stands with adjustable height permit easy manipulation of stock for marking, cutting, drilling, and welding.

The attachment of a tilting vise permits accurate location of angled holes. This model of a vise can be tilted in two dimensions.

An indexing attachment permits precise angular movement of the work piece. This model can be angled, as well as rotated.

Attaching this cross-slide vise permits very precise front-to-back and side-to-side location of the work piece under the drill bit.

times longer than regular high speed steel (HSS) bits. TiN bits can also be used in handheld drills.

MILLING MACHINE

The milling machine is similar in concept to a drill press but is much more precise, robust, and capable of more specialized work, such as precisely flattening the sealing surface of a valve body, cutting slots, or machining keyways. While the drill press is pretty much restricted to one job in metal, the milling machine can perform a wide range of specialized cutting to achieve various cuts and finishes in all kinds of materials, from steel to plastics.

With the influx of low-priced machine tools made overseas, plus conversion by industry to computer-controlled machinery, new and used conventional milling machines are becoming affordable and common in farm workshops. Some milling machines are integrated with metal lathes to make a combination machine tool.

METAL LATHES

By rapidly rotating a metal work piece held in the chuck and allowing precise placement of a cutting tool, the metal lathe can be used to create round shafts, taper the ends of bars, apply knurling to slightly increase the diameter of a worn-out rod, or many other workshop tasks. Alternatively, a cutting tool can be clamped in the chuck and rotated against a fixed work piece.

Until recently the conventional metal lathe was something that was found only in specialized machine shops. But as with the milling machine, recent industrial trends have made a lot of highly capable, affordable new and used machines available for private workshops. A certain amount of specialized operating knowledge is needed to make use of the lathe, but any library has good books explaining how to do it, and most mechanically apt people can pick up the basics fairly quickly. You can find books, such as *Metal Lathes: How to Run Them* by Fred H. Colvin and *Home*

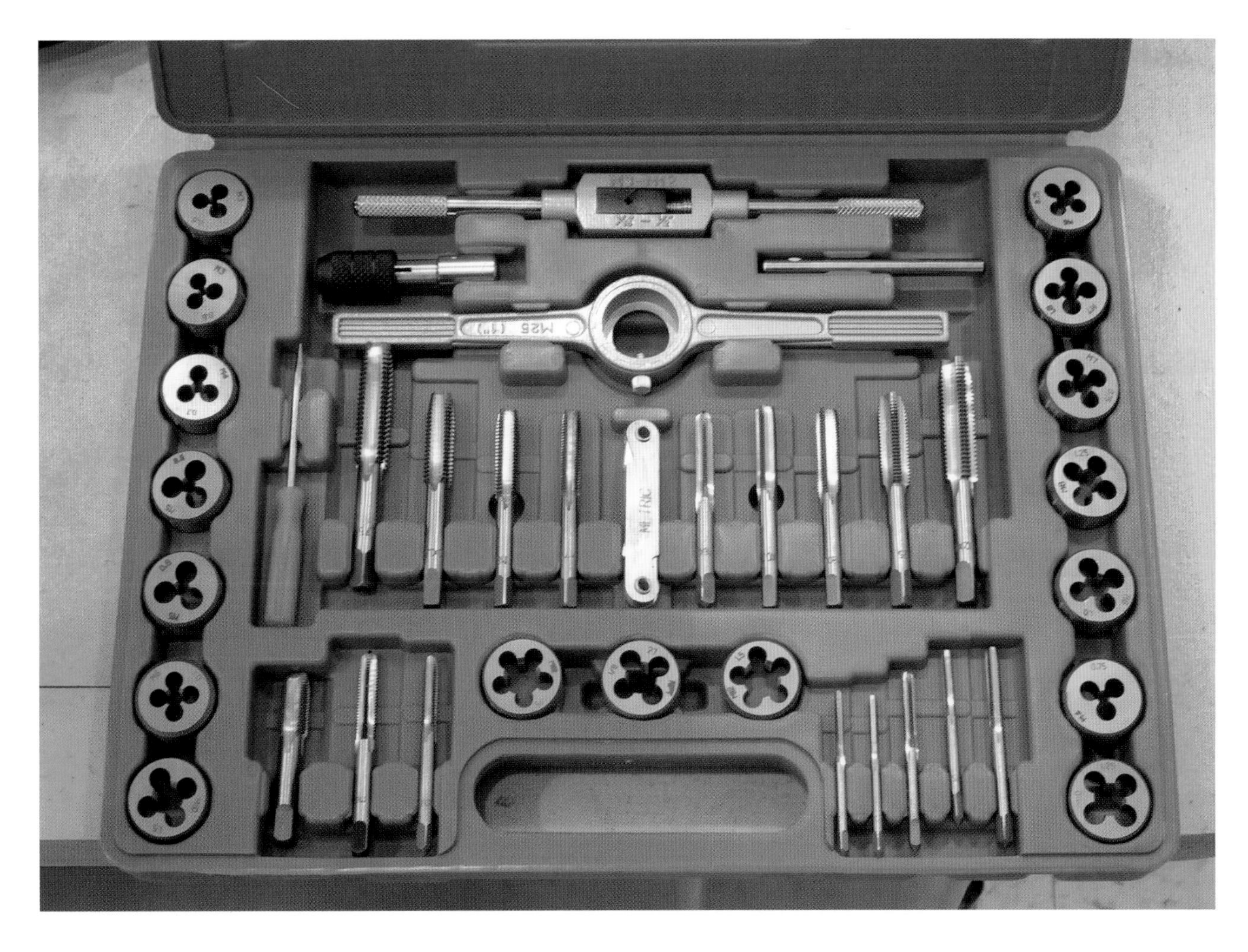

Taps and dies can either be bought as sets or individually to replace them as they break. The tough, hard steel is also quite brittle.

For any metal shaping job beyond a few strokes with a hand file or sandpaper, grinders can do it faster with equal or better precision.

Machinists Handbook by Doug Briney at bookstores and online suppliers such as Amazon.com.

BENDERS

One of the most interesting metalworking tools that modern technology has put within the reach of any farm workshop is the metal bender. These tools allow you to form complex shapes from metal tube, bar, or flat stock. A hand-powered or hydraulic bending machine lets you easily make pieces such as gas or fluid transfer pipes, handrails, gates, or racks.

The benefit of a bender is that it forms the bend without flattening a tube or letting the sides of solid stock bulge out. It also lets you make a bend without the need for heating, which could harden or soften the metal once cooled.

CHAPTER 12
SHEET METAL AREA

Farm workshops have ample opportunities for using sheet metal repair and fabrication equipment. Elsewhere, the equipment is most often associated with heating duct fabrication or auto body panel work, but on the farm the potential tasks go well beyond that.

Tractor sheet metal panels will get banged up in the course of use, especially when working around livestock. Bales, gates, walls, and animals themselves can crease the metal, or you may want to devote some hobby time restoring old tractors with wrinkled hoods, grilles, and fenders that can't easily be replaced any more. There is more reason than simple aesthetic appeal for fixing bent and crumpled metal. First of all is that the damage often interferes with efficient running of the tractor. For

Loader tractors get banged up pretty easily and fixing the tinwork is more than just an appearance issue.

Equipment without guards on the driveline is a bad accident waiting to happen.

example, a bashed-in grille increases the risk of the engine overheating due to chaff and straw blocking the spaces between radiator fins. Bent metal may interfere with access for service tasks, such as changing the air filter. Panels that no longer fit together well may allow rainwater to drip onto electrical wiring and cause corrosion that leads to poor starting and running. The second big reason to have the equipment for tractor sheet metal repair is that good external appearance is a big part of retaining resale value for the tractor.

Combines, especially older combines, are another farm machine where sheet metal equipment can be put to profitable use. The cutting, conveying, threshing, separating, straw handling, and unloading mechanisms of the combine are all housed in extensive areas of sheet metal that wears out in use. If you probe around the sheet panels on an older combine, you may find some places where the original thick sheet metal has worn to hardly more than the thickness of kitchen aluminum foil.

New sheet metal parts, if they are available at all, are expensive and often must be special ordered and result in extra downtime. Being able to fabricate a replacement panel in your shop can keep you running during harvest season and can also help keep older machines operating efficiently, longer. You may also be able to replace ordinary sheet steel panels with shop-built stainless steel equivalents that result in parts with improved wear resistance. As with the tractor, keeping the combine's sheet metal straight and strong protects your asset value by retaining more value at trade-in or sale time.

Another excellent reason to have and use sheet metal equipment in your farm workshop is to maintain the integrity of driveline guards on combines, haying machines, feed mills, augers, and other powered equipment. If left unguarded, chain drives, gear trains, and rotating shafts can

Even if parts are no longer available or too pricey for our tastes, it's possible to fabricate sheet metal substitutes.

quickly injure or kill an operator or bystander who accidentally comes in contact with the equipment. Straightening a bent guard or fabricating a replacement for a lost one can be a simple, satisfying job that pays incalculable safety dividends, especially if spouses, children, visitors, or animals are ever around the machines.

If you have livestock on your farm, sheet metal repairs will be a relatively common need because animals love to scratch themselves by rubbing against any strong object, such as buildings, machinery, and feeders. The edges of sheet metal panels provide a satisfying scratch, so they will get plenty of rubbing, which in turn means they eventually work loose and bend. If left long enough, they will completely break off and be trampled by sharp hooves into a crumpled mess. It becomes a sheet metal job and something your shop can easily and cheaply handle with the right equipment.

A sheet metal gauge is a type of fixed caliper with precisely milled gaps to assess the size of existing metal and confirm the size of metal to be formed. As the numerical gauge gets smaller, the thickness of the metal increases.

Finally, sheet metal makes a particularly good workbench top if you're planning a new bench for the shop. The metal resists oil staining and provides a flat surface that keeps parts from rolling around, plus the good light contrast makes parts easy to see. Since the required shapes are simple 90-degree bends, it is easy to have bench tops made at a commercial sheet metal shop. The trick is getting the formed tops to your shop without adding a few unintentional bends along the way due to bouncing around in the back of the truck. With a farm workshop that's equipped to make the bends, you can make up bench tops on the spot where you can save the cost of having it done commercially and get them into place while they still have the intended shape.

You may also find plenty of other uses for sheet metal equipment on the farm, such as making strong and weatherproof electrical switch and meter enclosures, grain bin aeration ductwork, welding area ventilation hoods, and more.

Fabricating new objects from sheet metal first requires a suitable layout area. For large pieces of sheet metal, working on the floor seems like a natural thing to do. However, the problem with using the floor as your layout area is that small rocks and holes can dent the metal as you lean or kneel on it. This may not matter in some cases, but for most projects it will prove a frustrating source of required extra smoothing work. If you must use the floor, make sure it is swept thoroughly, and then sweep a couple more times to make sure it is clean. Better yet, lay clean plywood down. Don't use soft pads such as cardboard or carpet on the floor, because the sheet metal will easily dent into the soft pad as you kneel or press on it with the heel of your hand.

Before drilling sheet metal, it's recommended to make a punch mark on the spot to be drilled. This centers the drill point in the hole and helps make sure it does not skitter off the mark. The automatic center punch, as used in the metalworking area (see Chapter 11), will also be very useful

Combines have many sheet metal parts that wear out due to abrasion from the crop or from rust. Replacing the parts with stainless steel solves both problems.

A sheet metal gauge is a type of fixed caliper with precisely milled gaps to assess the size of existing metal and confirm the size of metal to be formed. As the numerical gauge gets smaller, the thickness of the metal increases.

This combination shear, brake, and slip roller is for jobs involving relatively narrow pieces of sheet metal.

Grain-handling equipment often requires repair to sheet metal driveline guards or bin door panels.

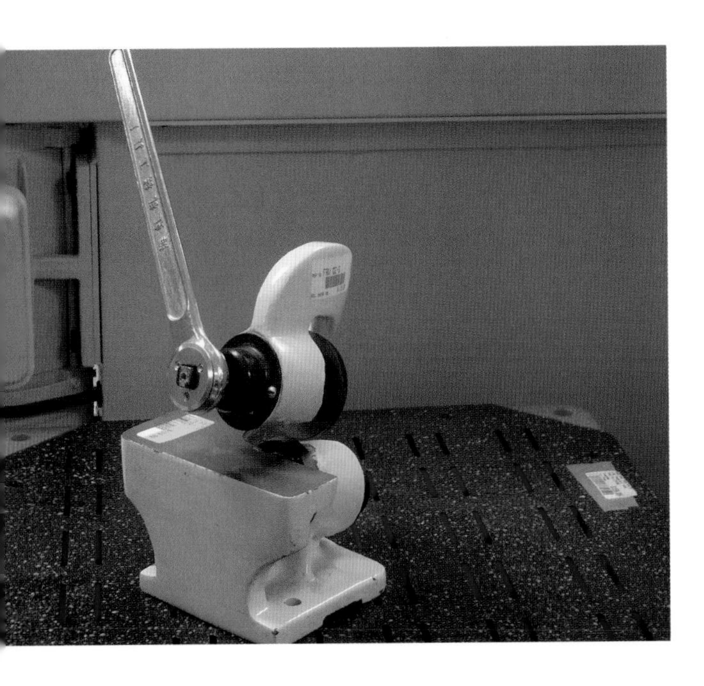

A ratcheting shear operates in a way similar to a can opener. Moving the lever advances the metal against the cutting wheel, permitting both straight and curved cuts to be made.

The foot-operated shear quickly and accurately slices large sheets of metal.

here. All you have to do is press the point down on the mark and at a preset pressure, the internal spring trips and drives a striker against the tip that punches a mark into the metal. With thin sheet metal it may even drive part way through the metal but that's all right.

Once a piece is cut, the sheet metal brake is used for quickly making straight-line bends in sheet metal. Many brakes also have a shear capability in order to allow quick and precise cutting of metal.

Aviation snips have handles that are color-coded: red for cuts that curve to the left, green for cuts that curve to the right, and yellow for straight cuts (the red and green snips can also make straight cuts when needed).

If you're just getting set up and don't have a brake yet, there is equipment that can come close to equaling the capabilities of the brake. It won't be as fast or as convenient, but it will get a lot of jobs done until a sheet metal brake can be obtained. The key is to have long jaws that clamp the metal at the required bend line. For short lengths of bend, a wide-jawed locking pliers may do the trick. All you need to do is clamp them on and bend the metal over by hand or with a hammer.

For longer distances, lengths of angle iron clamped along the bend line can make a fairly useful bending mechanism. It does get awkward once the length of the angle iron gets longer than twice the depth of your deepest clamp. At that point it is hard to obtain sufficient clamping force at the middle of the bend.

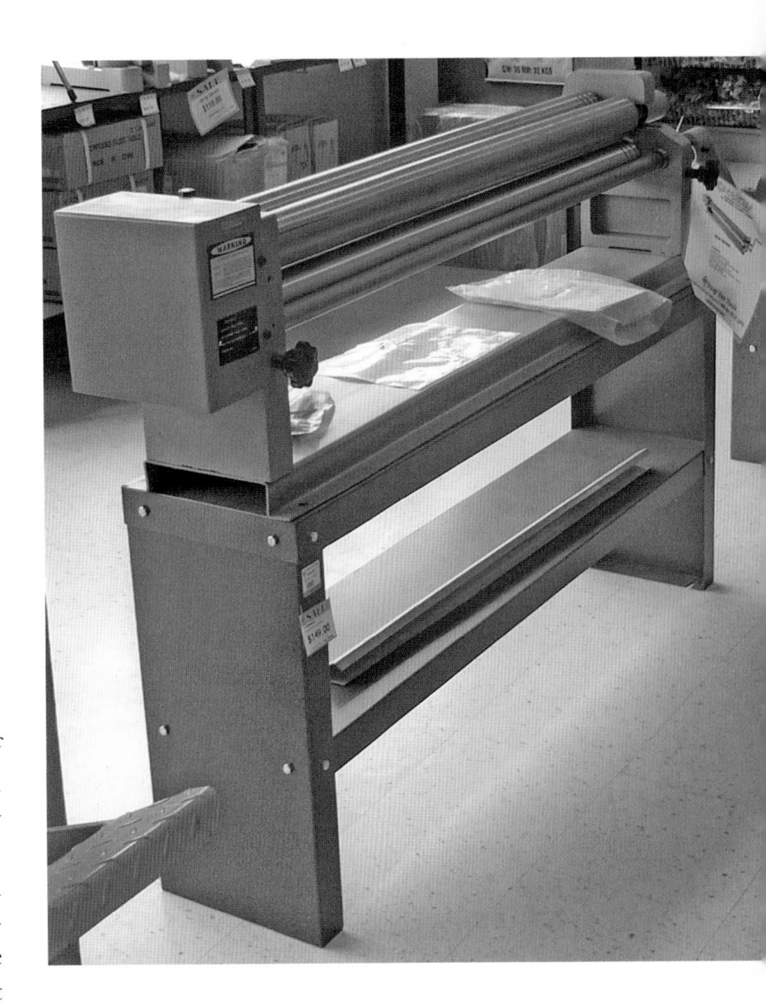

By passing metal over rollers at different heights, the slip roller permits curves to be pressed into the metal sheet.

For hand shearing, aviation snips speed up the work and improve accuracy.

Cutting with tinsnips can slowly, but surely, achieve almost any shape.

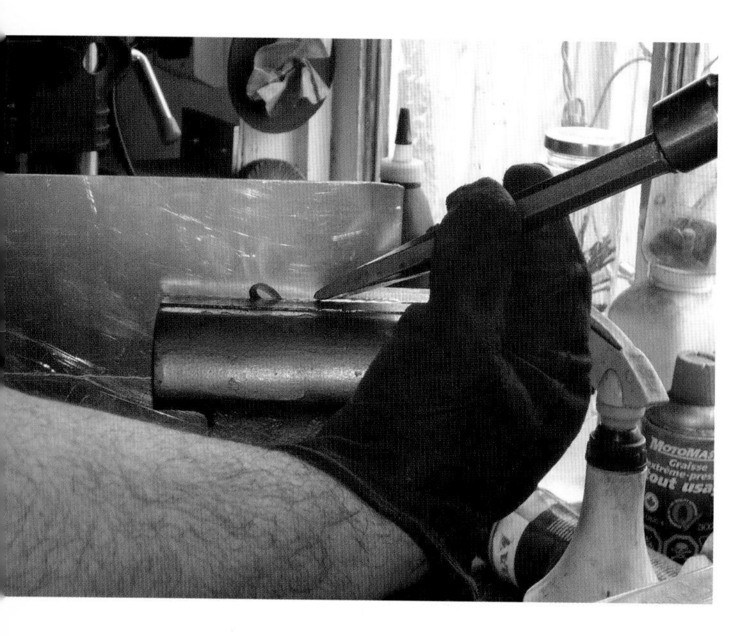

For hand-cutting thicker stock where hand snips prove too hard to operate, a chisel will shear through metal clamped in a vise.

If the non-bent part of the work piece is not too wide, you may be able to use your portable folding bench as the clamp, and angle-iron facings attached to the wooden jaws will facilitate the bend. Depending on the thickness of the metal, hand pressure or a wooden club can be used to push the protruding part down to form the bend. Running a heavy block of metal back and forth over the bend helps smooth it out. Countersunk screws can attach pieces of angle iron to the jaws of the portable folding workbench. The work is clamped with the part to be bent pointing up and the irons clamped on the bend line.

Forming simple curves such as O- or U-shapes can be done with a roller, which presses the middle of the curve down slightly as the work piece rolls along on wheels that support the two sides. If a sheet metal roller is still on order for your shop, you can alternatively form a slight bend in the metal over a piece of pipe or metal bar, then slide it slightly to the side and continue the process. As you gradually work all the way along the metal, the desired form will start to appear.

A plan for a unique workbench specialized for sheet metal bending, forming, and joining is included in the book *Working Sheet Metal* by David J. Gingery. If you're working on older tractors and vehicles where sheet metal work is needed, this type of bench can be a useful addition to your shop. The book also includes information on sheet-metal forming techniques based on the author's decades of experience in commercial shops and teaching sheet metal work.

COMPLEX BENDS

If you're planning on forming more complex bends in sheet metal, such as the shapes needed to make complete tractor hood and fenders or truck body components, things become more complex. Specialized sheet metal tools such as a planisher and English wheel will become potential candidates for your farm workshop equip-

Body worker's hammers and dollies are available in inexpensive sets and help you work out wrinkles and bends in any sheet metal.

ment. Since the topic is so well covered in aircraft and auto body sheet metal courses, books, and instructional videos/DVDs on hot rods and custom motorcycles, these resources should be consulted for acquiring the best equipment and necessary techniques required to effectively use the equipment.

A few useful references include:
Ultimate Sheet Metal Fabrication Book by Tim Remus

Traditional sheet metal forming explained with text and pictures:
www.artmetal.com/brambush/forging/proj06/index.html

Step-by-step guide to building your own English wheel:
www.roddingroundtable.com/tech/articles/12ewheel.html

Plans and parts to build your own planishing hammer:
www.tinmantech.com/html/little_powerhouse_air_hamm
er_p.php#plans

Step drills avoid the ragged edges left by using ordinary twist drills in sheet metal.

CHAPTER 13

EQUIPMENT FOR DRAINING FLUIDS

Various lubricants, coolants, hydraulic oils, and other fluids are necessary to farm equipment operation, and all of these fluids eventually have to be changed. Draining or removing the old fluids is generally not a task anyone really looks forward to or enjoys doing. The result is that a lot of essential service tends to get put off or even worse, overlooked entirely as long as the equipment appears to be running all right.

Having the right kinds of shop equipment that let you conveniently, easily, and quickly get rid of old fluids makes servicing jobs a whole lot less of a chore. This makes it much more likely that your equipment will receive the kind of service that helps keep it in top running order and results in increased value at trade-in or sale time. And although it's not strictly related, it tends in practice to have the benefit of resulting in a cleaner shop and yard. That's because with the right equipment to drain fluids, there are less dirty fluids slopped out on floors, yards, and mechanics. With the right draining equipment, it tends to be easier to get the old fluids into a holding container for disposal or recycling.

Whether it's for a normal oil change or occasional work involving less-frequently changed hydraulic system fluids, manual transmissions, and fixed-ratio gear cases, having the right kinds of draining equipment is an often overlooked way to make your shop more efficient and satisfying.

This rolling drainer is also equipped with a fitting (visible at lower left) that connects to a pump that transfers waste oil directly to a holding tank.

This inexpensive, but very efficient, oil drain pan is handy for transporting oil to disposal stations. When using the pan, make sure to move the funnel to the inlet, as shown. The center location is only for funnel storage and does not connect to the inside of the pan.

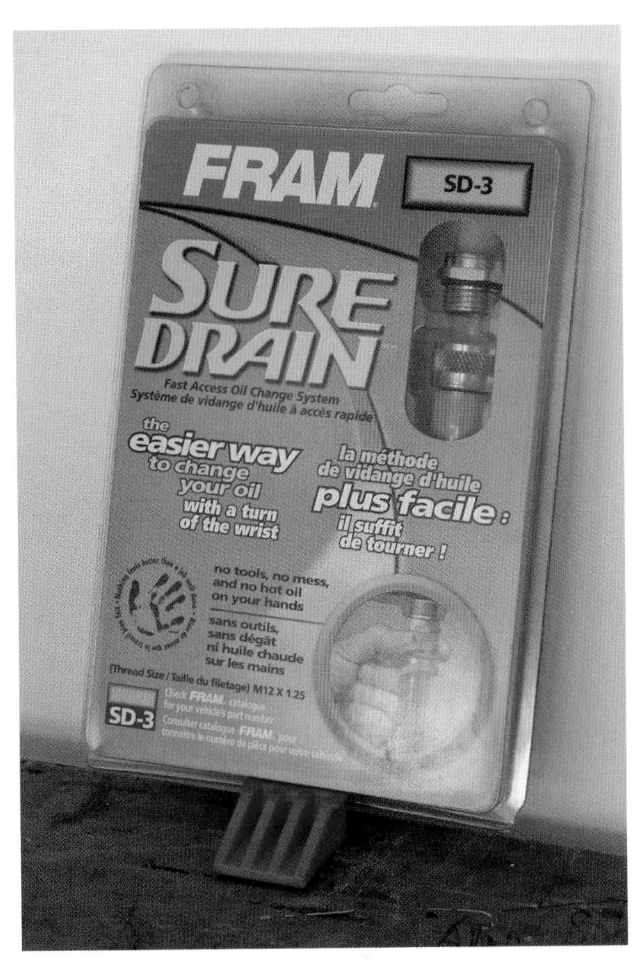

In the drain valve, this is the valve body that replaces the ordinary screw-in plug.

WHY FLUIDS BREAK DOWN

Lubricants, coolants, and hydraulic fluids eventually break down under the normal heat and pressure of use. The well-known need to change engine oil after a specified period of use is a prime example.

Another very important but much less appreciated reason fluids break down is through their accumulation of contaminants, especially water and acids. It's the reason why engine oil should be changed prior to putting away tractors and other powered equipment for the winter or for changing oil more often in a diesel engine that never really warms up to full operating temperature or idles a lot.

Crankcase oil in any engine always contains acids and water formed in the normal combustion process. In normal use when engines are hot and oil is changed regularly, these contaminants are not a problem. But if the contaminated engine oil is allowed to just sit over the winter months, the different specific gravity of oil and water makes them separate. When that occurs, concentrated water and acids may gradually corrode, gum up, or form scale on hard-to-reach engine parts, such as main bearings.

A diesel engine that operates at cooler-than-normal temperatures and under light load forms much higher-than-normal levels of contaminants that wind up in the engine oil. So while a lot of idling and light driving around the yard might put little load on the tractor's metal parts, it also leads to more contaminants in the engine oil.

In addition to breakdown under stresses of use, the job-specific polymer chains in coolants and automatic transmission fluids also break down over time. Brake fluid is also well known for becoming contaminated with water from the air, which reduces performance and necessitates periodic changing of fluid.

DRAINING, LARGE AND SMALL

Equipment for draining includes drain pans for equipment with gravity drains (engine oil pan plugs, radiators) and suction devices for reservoirs without a drain plug.

There are good reasons to use drain equipment somewhat more sophisticated than an ordinary bucket or old ice cream pail. More specialized equipment has generally been designed to eliminate some of the common frustrations users have encountered with other equipment. For example, a specialized oil drain pan is low and shallow with a wide opening and short lips extending inward from each side at the top. The low height keeps it from getting in your way while you're crawling under the machine. The shallow depth makes it easier to find the drain plug when it falls into the dirty oil being drained. In some large drain pans,

The drain hose attaches to the valve body. When the hose is attached, the oil drains out. Flow can be halted if your disposal container gets full partway through.

a screen at the inlet accomplishes that task even better, and provides a method of letting the oil filter drain out, as well.

The extra-wide opening of a specialized drain pan makes it more certain that the pan is well positioned to catch all the oil as it drains. Often as engine oil gets down to the final thin stream, surface tension will make it travel part way along the engine oil pan before gravity takes over to make it fall straight down. Unless the pan has a wide opening, that often makes the oil drip out beside, not in, the drain bucket. Large drain pans on caster wheels make it easy to roll the pan into position under a tractor and back out. Finally, the lips on each side keep the oil from slopping over the sides of the pan as you move the full pan out. All of the features provide benefits over using any bucket or plastic ice cream pail, especially when working with hot, dirty engine oil.

Specialized oil drain taps can also make the oil-draining job considerably easier and less messy. The standard threaded plug in the oil drain point is replaced with a spring-loaded valve of the same thread size.

A manually pumped siphon tank provides a considerable improvement over converted grease guns. Note the variety of suction wands supplied in the long black holder at left.

When the protective cap is removed and the second part of the drain tap is attached to the drain valve, the spring-loaded valve is pushed open so the oil can flow out. These specialized oil drain valves cost from $20 to $50, depending on manufacturer and application, and offer two big advantages in use. First, they eliminate hand contact with hot, dirty oil; and second, the hose attachment on the drain tap lets you direct waste oil straight to a disposal container. This makes the unit particularly handy for use in combines and other equipment where the engine is placed high up on the machine. Being able to direct the old oil straight into a disposal container saves you from having to rig up a temporary drain trough or hose to get old oil down to

Veterinary syringes make handy suction or filling devices for small fluid reservoirs.

ground level. These automatic drain taps are available at most tractor or automotive parts suppliers or online from www.fumotovalve.com.

For getting fluid out from reservoirs that don't have a drain plug to unscrew, you will need equipment to apply suction to remove fluids through the fill hole. There are simple manual oil suction guns available from oil and tractor parts suppliers, but get advice from staff when choosing one. Many have problems with fluid rapidly drooling back out as you move the suction hose from the reservoir to the disposal container, which creates a mess. This problem seems to be unavoidably related to the large size of the suction piston so ask the staff if the one they sell works well. For large reservoirs, a better alternative for the farm workshop is one of the newer generations of suction tanks that are operated by compressed air or a manual-charging handle.

For small reservoirs, a well-functioning option is a large veterinary syringe with a length of hose attached to the opening where the needle would ordinarily attach. These syringes, complete with a hard protective case, are available for only a few dollars at farm supply stores, and a few feet of clear tubing will run a dollar or two more. Since syringe bodies must develop reliable, easy-to-feel pressure with zero air leaks in their intended use for injecting animal medicines, they also develop tremendous suction when removing oil or other fluids from a reservoir. The largest size removes less than 100 cc (4 ounces) at a time, but removal is quite rapid. The strong, reliable suction also makes these syringes great for positive suction of air bubbles out of brake systems, which eliminates a lot of the mess involved with ordinary bleeding procedures for paint-destroying brake fluid.

DISPOSAL OF DRAINED FLUIDS

Dumping waste fluids on the ground is not an option, even out in the country far from the prying eyes of regulators. With oils, some maintain that since oil originally came out of the ground, there's nothing wrong with pouring it back in the ground to settle the dust or burying it in a shallow

Waste oil holding tanks can be located outside a farm workshop. The line leading through the wall goes into the workshop, where it can connect directly to fittings on oil drain pans, such as the one shown in the first photo in this section.

hole. There are many things wrong with that opinion. Oil generally comes from many thousands of feet down below the topsoil. The oil you use for lubrication is not the same as what came out of the ground. Engine oil contains many very useful lubrication-enhancing additives, and these are not the kind of thing you want in the topsoil and water that's going into your farm products.

It's getting to be common that an environmental audit may be required for any commercial real estate transaction. Farms such as yours could be considered commercial real estate, perhaps not in your jurisdiction today, but can you be sure it will be the case about 10 or 20 years from now? If your yard is polluted with oily soil, leaking batteries, and so on, you could be faced with a huge topsoil remediation expense before you can sell your land.

Why take the risk, especially when there are so many free fluid disposal and recycling facilities now available at farm fuel and oil suppliers, county waste disposal sites, and fire and EMS stations? These days, environmentally minded disposal is both easy and much to your advantage as a way to protect the value of your farm.

These disposal sites usually also offer tanks for old coolant (antifreeze) so you can safely dispose it at the same time. Coolants are not only toxic, but have a sweet taste that makes them inadvertently attractive to children, livestock, pets, and wildlife.

CHAPTER 14
MEASURING AND REFILLING

Efficient service and long life of machines depend on a sufficient source of the right fluid, whether that is a lubricant, hydraulic/brake fluid, coolant, or some other type of fluid such as windshield washer fluid. An essential part of the farm workshop is equipment to get the right amounts of the right fluids into the appropriate reservoirs without spillage. None of the fluids you're putting into farm equipment can be thought of as cheap anymore, even windshield washer fluid. Having part of it go to waste by splashing around as you try to hit a small reservoir opening can be permanently eliminated with a few dollars worth of investment in simple and inexpensive equipment.

Since many of the fluid fill holes on tractors, implements, trucks, and other equipment are quite often in awkward-to-reach, hard-to-see locations, this kind of equipment will help finish your shop tasks a whole lot easier and leave you feeling less stressed.

Spilling not only wastes your money and your time in cleaning up the mess, but it also makes the fluid level inexact. That may be fine where you have a dipstick to check level and can add or subtract a little bit, but some reservoirs don't have that. One common example is the differential oil reservoir in the rear axle of Ford trucks of the 1970s and 1980s. When the correct measure amount of oil is put in, the oil level is a small specified distance below the filler plug level. If too much oil is put in and is up to the level of the plug threads, the resultant pressure that builds up will blow out the oil seals at both ends of

Taking care of fluid levels in expensive modern farm equipment is no place for eyeballing measurements or spilling fluids that attract dirt.

Whether you prefer plastic or metal, it's easy to get good-quality calibrated jugs to eliminate measurement guesswork.

the axle. This leads to a considerable mess, bother, and expense to replace the seals. All that can be avoided by having the equipment to measure the right fluid and get it in the reservoir without spillage.

FILL 'ER UP, MAC

When it's time to fill a fluid reservoir back up, a few inexpensive pieces of shop equipment are very handy unless the reservoir is designed to hold a whole number of quarts, gallons, or liters. Properly calibrated measuring jugs are easily available and should be part of every shop's equipment. For about $10, they are a simple way to eliminate guesswork and estimation involved with makeshift containers.

Measuring jugs need to have quantity markings clearly stamped into or molded onto the sides of the vessel. For plastic containers, the clearer the body material of a measuring container, the easier it is to see the fluid level while filling up with fluids that are not dark enough to provide good contrast in low light. This issue of clarity and contrast tends to get worse as the body of the jug becomes stained from exposure to various liquids and to ultraviolet light.

This funnel looks like something from the steam age, but it's still a very useful piece of equipment, especially when it includes a screen in the outlet.

The awkward location of many fluid reservoirs makes the job of refilling much easier if an offset funnel is available.

When filling the jug, provide a flat, level surface so that you can be certain the level is correctly indicated.

When choosing fluid measuring jugs, look for something with a well-defined pouring spout so that fluid doesn't dribble down the sides when pouring out. A completely enclosed spout will pour slower but more accurate than a spout with a partly or fully open top.

For getting fluids from the jug or bottle into the reservoir, a variety of inexpensive funnels and bottle attachments make the job more efficient and less messy. A simple standard conical funnel is often surprisingly ineffective due to its design, which probably comes from the days of transferring fluid from small milking buckets into a larger pail. When this type of wide-mouthed, long-stemmed funnel is used to put fluid into engines and equipment, conditions often prevent the long outflow stem from completely entering the reservoir. For example, a valve gear or oil spray baffle may keep the spout from fully entering the engine oil filler hole. If that happens, one hand is needed to hold the funnel to keep it from sitting crookedly or else the funnel will tip when filled, spilling fluid everywhere except where it's supposed to go. Better equipment is needed, and fortunately better designs are easily available at low cost.

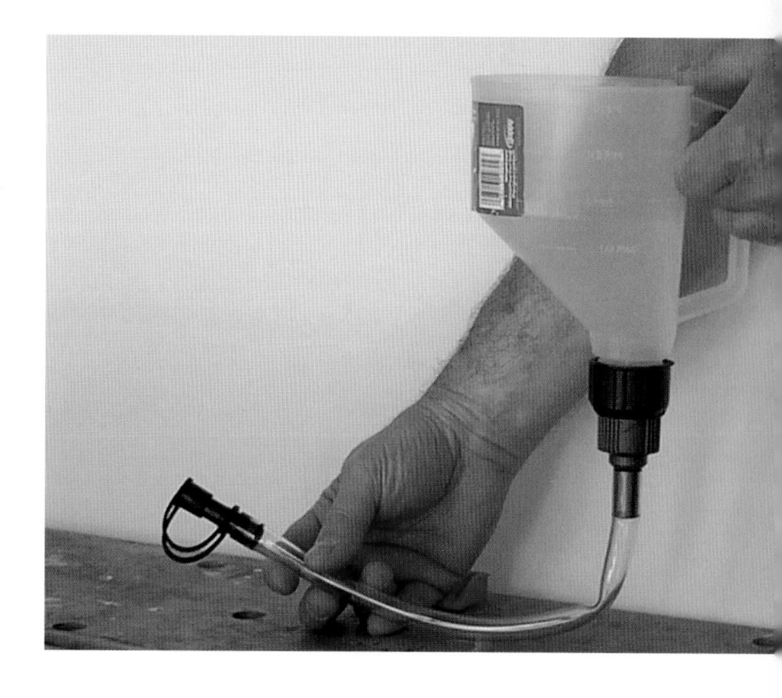

This funnel design combines a measuring body, an on-off valve in the outlet, and a long flexible hose to reach pretty well anywhere. The fluid does flow rather slowly, so if you're filling a large reservoir, be prepared for a long job.

The basic screw-on spout for quart bottles eliminates the spillage that always seems to occur just as you almost get the bottle mouth in the filler hole. Twisting the spout can halt the flow.

One of the oldest improvements was to put a long semi-flexible hose on the bottom of the funnel to make the traditional gooseneck arrangement. The key is that the semi-flexible metal hose holds the curve you bend into it. This allows the funnel to be set so it can be leaned or braced against something atop or beside the reservoir. When braced steadily in position while it's filled, the curved spout flows the fluid into the reservoir. For reservoirs that may be awkward to fill because of machinery above the opening, it's often possible to slide the spout several inches into the filler hole, then bend the tube so the funnel opening is firmly held up offset from the hole for easy filling. The metal gooseneck funnel is an old design but still provides enough benefits to keep it useful in modern shops.

A modern derivative uses a plastic flexible spout that is equally bendable but springs back into the original shape once released. To get over that disadvantage, many of these designs place the outlet at one end of the funnel. This helps keep it balanced, in place, and reduces the tipping force as

Pictured here are bulk oil dispensing barrels. Note the use of hair covers to keep dust and insects out of the measuring jugs.

Cordless electric grease guns make greasing much less of a chore because it frees one hand for guiding the hose to awkwardly placed fittings. The easier it is to grease the machine, the more likely it will properly get done.

the funnel body is filled. The plastic bodies of these funnels are also rather more resistant to bending from impacts, although a question must be raised as how a funnel could possibly be used that results in it suffering heavy impacts. That aside, offset plastic funnels have a definite place in the fluid dispensing area of the farm workshop.

KEEP THEM ALL CLEAN

Your shop's fluid measuring and dispensing equipment must be kept clean to avoid introducing contaminants into engines and equipment. If refilling jugs and funnels are left sitting on a shelf or on top of the oil drums, they will attract dust and trap insects. This mess can be cleaned off before use, but it just adds to the work and makes your shop less efficient.

A better solution is to remove the oil or coolant film by quickly washing the container in solvent or water, and then store it in a closed, dust- and insect-proof cupboard. This keeps your fluid measuring and dispensing equipment in good condition and ready to use when you need it.

Stick-on service reminders keep the maintenance reminder right where you can see it during equipment operation and add a touch of professionalism that comes in handy at trade-in time to help show the equipment was well maintained.

CHAPTER 15

INFORMATION MANAGEMENT SYSTEM

Do you find it irritating when you're in the middle of working on something and the phone rings with a call you don't want to miss? Have you ever had a job delayed because you couldn't find the manual that tells how much oil to use in some piece of equipment? Ever thought how nice it would be to be able to look at equipment parts books online or print out a page to use as a guide to reassembly? If you answered "yes" to any of these questions, then you need equipment for improved information management. If you're planning a new shop and foresee any situations like that coming up, it's worth taking some time to build in information management features and practices that will make your shop a more efficient and satisfying place to work.

Once you've worked your way into a position like this (necessary to service a machine), you're not going to be too happy about needing to get out to look up a specification or answer the phone.

To be useful, manuals need to be where you can find them and probably in some kind of order other than geological (i.e., piled in the order they were last laid down).

#@%& PHONE

Seems like it never fails. You just get deeply involved in some workshop task and the phone rings. It might be just a conversational call from a friend or a telemarketing call, but you never know. It could just as easily be the dealer returning your call on whether the part is available or a customer wanting to buy some of your farm's products. But by the time you listen to make sure it was the phone, extract yourself from the task, lay down the tools, wipe your hands, and get to the phone, it stops ringing.

If service coverage is good in your area and you don't mind any extra minute charges that may be involved, carrying a cell phone is a great alternative. But in practice, high battery–power consumption and their annoying tendency to fall into dirt and oil puts some limitation on cell phone usage; and a cell phone usually doesn't handle calls that come to the home number.

The first step to better call handling on a farm landline is to make sure you can clearly hear the ringing. Most telephone service providers to rural areas can supply, at low cost, a remote ringer. This loud little bell relays the indoor telephone ring tones out to your yard, shop, or other outbuildings. This auxiliary phone ringer can be handy even if you do have an auxiliary phone outlet in the farm workshop.

With an auxiliary outlet in your farm workshop, you can make excellent use of one of the new 2.4 GHz or better cordless phones. The range and clarity of these units means it's easy to keep one in a pocket of your coveralls and have

It's perhaps not the best place for manuals, but may be all right if only a few pieces of information are involved.

it relay calls cleanly even if you're underneath or inside some large metal object such as a combine or grain bin. If you add a hands-free headset, perhaps even a wireless headset, it creates a great phone information management option for the farm workshop.

MANUALS

Among other useful information, the operator's manuals, that are included with new equipment also contain specifications on service, such as tire pressures for optimum traction, how much engine oil is required, and the torque to be applied to the oil drain plug upon reinstallation. The problem that happens in almost every shop is that since very few people actually read these manuals (sad but true), they tend to get tossed in a drawer or piled on a shelf with no particular form of organization.

The result is when it's time to consult the service specifications, the mechanic may waste time rooting through the entire stack of manuals before finding the correct one. The information may also be ruined by a factor unique to farm workshops: many mice live in the country and love to chew books and make nests in paper. There's nothing like having a big hole where the information used to be! For manual storage that wastes less of your time and remains useful, keep them in a clearly defined place where your shop cat(s) can easily patrol, and insert simple cardboard dividers for various types of equipment (e.g., tractors, pump engines, implements). Another suggestion is to put them in a plastic container found at any discount store. It will keep the manuals protected from mice and water damage.

Another useful information management device is to extract key items of information and keep a durable copy where it will be needed most often. For example, service specifications such as engine oil quantity or bolt torque values can be written on a card and laminated at a very low cost at an office supplies outlet. Information on service intervals can be relayed via a simple window sticker or tag.

REASSEMBLY GUIDES

When taking things apart, it always seems easy to remember how it goes back together, even if there are different lengths of bolts involved in the assembly. But then you clean the components in the parts washer, perhaps even do some welding or metal shaping. Even more likely, you leave the job while you wait for parts or attend to one of the many urgent tasks that the farm requires. Then, on returning to the task of reassembling the machine, there suddenly seem to be severe information gaps on how it goes back together.

Providing yourself with a few simple reassembly guides is a way to get past the limitations of memory and free your mind for more important things such as planning a livestock marketing strategy or remembering your wife's birthday.

Photos, sketches, or notes taken during disassembly are very valuable but are often difficult to do when working on heavy, dirty, greasy equipment. In practice, having to clean off your hands before you can make a sketch or use a camera tends to inhibit creating good records of disassembly. A useful alternative is to use large pieces of thick cardboard where you punch holes with a nail to hold bolts in the

Another approach is to leave the manuals right at the machine they refer to, which is relatively workable for shop equipment that stays indoors.

pattern similar to what they had during disassembly. Additional notes can be made on the cardboard with a felt marker, which is easier to hold with dirty hands or when wearing mechanic's gloves. The cardboard also soaks up oil and grease to help keep the work area clean.

If there is no manual available to show how parts go back together, it may be possible during working hours to request a fax from the dealer of the parts diagram for the area you're working on. The exploded drawing gives a good idea of the relationship of one part to another and what kind and size of fastener holds it together.

They may look lazy, but these efficient little killers help keep mice from making nests in your operator's manuals. Female cats are generally much more enthusiastic hunters.

CHAPTER 16
SHOP SAFETY

Building a farm workshop with the tools and equipment you need can be a very satisfying experience, not only in terms of being able to keep the farm running efficiently and economically, but in terms of having a showcase for your hard-won competency and craftsmanship. It would be a sad thing if your shop turned out to be a place that created the sorrow and pain of injury to yourself or a family member. The same equipment that allows you to accomplish so much can, in the blink of an eye, turn into equipment that slices, dices, and crushes your future in many agonizing and expensive ways.

You can take control of your and your visitors' safety by making the decision to adopt safe work practices and use them consistently. Relying on "the way granpappy done it" may have some merit, but it may just as easily be something ol' granpappy would cuff you upside the head for not updating in light of new evidence he would have had the sense to take into account. Always keep up with learning about safe work

Price is no argument for restricting your shop for safety gear. Keep appropriate items near every piece of equipment. The easier it is to see and reach, the more likely it will be used.

Fire extinguishers and signs indicating their location should be near every door, because that's where you may rush in looking for one.

practices—it's your hide, not your pride, that's on the line. Make the decision to adopt safe practices, use them consistently, and make sure everyone in you shop does, too.

"Thinking safety" doesn't need to be looked at as weakness, timidity, or lack of manly boldness. It can just as easily and logically be thought of as taking pride in doing things right. Take a minute to imagine how you would react to someone using your tools or shop equipment for the wrong purpose. For example, suppose your neighbor came in and used the torque wrench to hammer nails and light his cigarette with the oxyacetylene torch while slightly drunk. You might understandably be moved to forcefully point out the error of his ways. Lay out a detailed program of how you intend to deal with any similar behavior in the future.

You can use this same natural, commonsense approach to developing and implementing a program for safety in the farm workshop. Instead of looking at shop safety as a tedious list of things to do and not do, look at it as using the shop tools and equipment the right way. Shop safety is taking sensible precautions and making sensible changes in methods of work to help make sure that your farm workshop equipment achieves the right purposes.

Right way: The equipment in your farm workshop is there to exert torque, pressure, cutting power, chemical energy, and so on in ways that make more things possible to accomplish efficiently and correctly.

Safe = Equipment and tools used the right way.

Wrong way: You have not spent good money and time getting workshop equipment to slice skin, tear muscles, break bones, burn eyeballs, poison you with fumes, or set the whole place ablaze.

Unsafe = Equipment and tools used the wrong way.

Safety starts with finding out the right way to use equipment. Consult with suppliers, read shop-related publications to stay current, take courses, and do anything else you can to satisfy yourself that you have valid information.

Become your own safety supervisor by following that up with constant, detailed observation on what works and what causes safety problems. Large commercial fabrication and maintenance shops in all industries know accidents cause lost productivity and increase training costs, as well as pain and suffering, so they have programs and training to keep the shop operating without incidents. In the farm workshop you are in charge, so one of the hats you need to wear is that of safety supervisor. Your attitudes and practices affect not only the risk to your own life and limb, but set the standard for others, including that nervy nail-hammering neighbor.

GENERAL SAFETY METHODS

When you acquire equipment, insist on learning how to use it properly. Don't just bring it home, plug it in, and find out by accident.

- Read the manual and learn any special operating techniques. This keeps you safer and improves productivity.
- Wear the right protective gear, especially eye protection. A tiny sliver of contaminated steel can ruin your vision for life and literally blind you to so much life has to offer. Consider the secondary consequences of injury, such as contemplating how hard it will be to herd your livestock if neglecting to wear steel-toed boots leads to a crippling foot injury in the workshop.
- Keep appropriate protective gear close to where it will be used so that it becomes no trouble at all to put it on.

KEY SAFETY TIPS

• Keep the shop clean. Metal scraps, cords, empty boxes, and scrap can trip you or lead to tools and work accidentally being upset or crashing to the floor.
• Avoid loose-fitting clothing, dangling hair, or jewelry that can get tangled in workshop equipment.
• Wear appropriate safety gear and use hearing protectors on noisy jobs.
• Protect your eyes. Metal chips are razor sharp and hard to remove, and welding slag can permanently burn eye tissue.
• Exercise extreme care when working on unfamiliar materials. For example, magnesium chips burn with great intensity under certain conditions. Chips and shavings from materials that contain asbestos, fiberglass, beryllium, or beryllium copper can lead to respiratory difficulties or skin problems.
• Horseplay, daydreaming, and other losses of concentration increase your risk of injury.
• Use machine assists or round up some helpers when you have to move heavy machine parts or large pieces of metal. Back injuries and muscle strains can easily turn into debilitating long-term injuries.
• If temporarily employing helpers such as your wife or children, don't expect them to know the job as well as you or to be able to lift as much. Lack of patience or respectful explanation, or exposing them to tasks beyond their skill and strength can lead to extreme danger you may forever regret.
• Only operate equipment when all guards are in place. If guards are missing, find replacements or fabricate some.
• Stop the machine to make adjustments or measurements.
• Never attempt to remove metal chips or cuttings with your hands or while the machine is operating.
• Chips produced by metalworking operations, such as threading, are extremely sharp so use a brush, stick, or piece of cloth, not your hand, to push them away. Avoid the use of compressed air because the resultant fast-flying chips may injure you or a nearby person.
• After cutting metal, remove sharp edges and burrs before moving on to the next step. These edges and burrs are as sharp as the edges of a tin can lid and can slice you just as severely.
• Always remove the key from the chuck before turning on the drill press. If it is not removed, it will tangle up in something or fly out with considerable force when the spindle starts to spin.
• Never disable dead man switches on grinders and other tools.
• Clamp down work pieces when using drill presses.

Keep equipment lubed, adjusted, and maintained. Poorly maintained equipment is more likely to expose you to danger. For example, a drill bit that's lubed while cutting is less likely to jam in the hole and make the work piece spin.

Practice assessing what could go wrong and take steps to avoid the resultant risks. For example, when you're cutting metal, think about where the hot slag might land and what would happen when it got there.

Take the time to make sure something is correct. Don't guess whether the clamp is tight or the axle is blocked, take a second or two to make sure.

THE FRUSTRATION FACTOR

Many farm machinery injuries are related to blockages and breakdowns that occur during use, such as clearing hay from an improperly functioning baler. When these blockages and breakdowns occur, the operator is likely to be in a frustrated or agitated frame of mind. This is especially so if the problem is repeatedly occurring or has not been resolved by normal measures.

When frustration builds up, reasoned judgment often goes out the window. Alternatives that are less risky but might take a little longer are overlooked or quickly rejected, even though the delay they bring is much less than the delay entailed by an injury. A person who feels the job is way behind now may give little thought to how much further behind things would be if he or she was injured.

Understanding that stress and frustration disrupt your usual thought processes is a reminder to have countermeasures prepared in advance. It can be thought of as similar

A firefighting cart with large wheels can be easily rolled over rough ground to wherever it's needed on the farm.

to the way pilots have a definite checklist to go through when they encounter a stressful situation. Farm stress checklists can be as simple as taking a deep breath, taking 30 seconds to consider the options before taking action, or making a commitment to always shut off the machine before making repairs or adjustments. When you're frustrated, thinking "just get 'er done" can more easily lead to a work-stopping or even life-ending accident. Having a personal checklist of countermeasures in times of frustration can be an alternative that improves personal safety in both the short- and long-term.

STARTING ENGINES

Farm safety records clearly indicate that among adults, bypass starting (operator standing on the ground during the start) is the leading cause of tractor runover accidents. On tractors with bad wiring, broken starter switches, weak batteries, or starters and/or missing or bypassed safety interlocks, many operators are tempted to work around these difficulties by jump-starting or short-circuiting part of the starter wiring. Starting the tractor in this way can result in the tractor starting while in gear and running over the operator or bystanders or punching a tractor-sized hole in the wall or door of the shop.

This farm also uses a surplus airport-type fire truck. With combines costing hundreds of thousands of dollars and the added risk of losing crops to fire, it could quickly pay for itself.

The sticker says it all. Follow the precautions to avoid painful steam burns

Until a proper repair can be done on the starter switch, all this danger can easily be avoided by making use of a remote starter switch available at any tractor or auto parts supplier for a few dollars. The wires of the remote starter switch clip onto the starter terminals. The length of the remote starter switch wires lets you stand clear of the wheels or sit in the operator's seat. When starting or boosting equipment in the workshop:

- Disengage the clutch (on manual gear shift tractors).
- Take the machine out of gear or place the range lever in PARK or START.
- If booster cables are used, connect them to the dead battery, not directly to the starter.
- Start the engine from the operator's seat with the starter control or a remote starter switch.

For more details on the dangers of remote starting and use of a remote starter switch, please refer to the *Farm Satety Handbook.*

> A 51-year-old farm worker died from injuries received when he was run over by the farm tractor he was operating. Because a key had previously been broken off in the ignition switch and the switch had not been repaired, to start the tractor the driver used a screwdriver to create an electrical short between the tractor's battery and starter, an operation that required the worker to position himself directly in front of the inner right front wheel while reaching into the engine compartment. The tractor lurched forward and ran over the driver, knocked down a bystander, and veered toward a roadway before being stopped by another person. The medical examiner listed the immediate cause of death as multiple head and trunk injuries.
>
> **from Oklahoma FACE Report #01-OK014-01**

BLOCKING FOR LIFE

Jacks, hoists, and hydraulic arms are made for lifting things and should not be counted on for holding up things once lifted: Do not rely on lifts alone for support. Many farmers have been killed or injured by making this mistake. Use mechanically locked jack stands or engage any mechanical locks on the jack or lift if it is so equipped.

A selection of stout wooden blocks or a vertical section of log is often enough, so keep plenty around to suit whatever needs for blocking may arise. If you do use a section of log, remember that a falling machine with any downward-pointing metal protrusions could split the log the same way an ax would. To help prevent this, lay a piece of flat lumber on top of the log. Any falling metal that strikes it will then be hitting cross grain, which is a lot harder to split.

SIGNS OF SUBSTANCE DANGER

Hazards on products used in the farm workshop are clearly marked with certain symbols. The Workplace Hazardous Materials Information System (WHMIS) is an international system of symbols that provides health and safety information about controlled products, such as solvents and extremely flammable materials.

Take a look at the product label for any special symbols and use the appropriate measures for working safely in the

Tripping over electrical cords is frustrating, embarrassing, and dangerous. Cord reels make it simple to remove that hazard from your shop.

An Iowa farmer was killed while working on a rotary mower in his machine shed. Using the hydraulics, he raised the mower to a sufficient height to work underneath but did not provide support or blocking. At one point he apparently tried to roll out from under the mower but became trapped between the right rear wheel of the mower and the mower deck. The mower continued to come down and pinned him to the floor, crushing him in the chest. The farmer was found dead under the mower in his machine shed.

from Iowa FACE Report #98IA56

presence of the marked material. For more complete information, inquire at the place that sold the product, or do an Internet search for WHMIS symbols.

While WHMIS labels only apply to controlled products, labels on consumer products may also have symbols that denote poisonous, corrosive, flammable, and explosive hazards. Take a moment to look at the symbol to assess the level and kind of hazard associated with the product. The symbols are designed to clearly communicate the type of danger and are usually accompanied by text that further explains it.

Labels for consumer (restricted) products (e.g., pesticides) have symbols similar to the WHMIS symbols above, contained inside an inverted triangle, diamond, or octagon, depending on the degree of hazard.

Symbol	Signal Word	Level of Hazard
Octagon	DANGER	High
Diamond	WARNING	Moderate
Triangle	CAUTION	Low

A symbol may also be included inside the octagon, diamond, or triangle hazard symbol. For example, the skull and crossbones symbol inside an octagon with the signal words "DANGER POISON" indicates a high poison hazard.

If you know the chemical or trade name of a product and want to get complete information on any potential hazards, you can easily find Material Safety Data Sheets (MSDSs) online for anything from acetone to xylene. Simply do an Internet search for "MSDS" and the material.

For agricultural chemicals such as pesticides, the product label and MSDS for every product sold in North America can be viewed free online at www.cdms.net. The product label contains key information on safety and usage and should always be consulted whenever these products are applied, handled, stored, or transported.

If a workpiece jams against the grinder wheel, it can result in several painful or costly consequences. Keep the gap between wheels and rests properly adjusted to minimize hazards.

COMPRESSED AIR SAFETY

While compressed air is tremendously useful for many shop tasks (see Chapter 3), it is tremendously dangerous when handled the wrong way. In particular, using high-pressure air for cleaning dust off your clothes is a very dangerous practice. Compressed air that enters the blood-stream through a break in the skin or through a body opening can cause an air bubble in the bloodstream (embolism) and lead to coma, paralysis, or death. Horse-play with the air hose has been a cause of some serious workplace accidents. Using compressed air to blow dust off equipment or shavings off metalwork causes those particles to blast into your eyes. A brush and shop vacuum cleaner is a better alternative.

The tremendous amount of energy stored in compressed air can lead to serious accidents or fatalities if overfilling a tire bursts the rubber or makes a split-rim band pop off the wheel. Since one tends to be leaning over or squatting quite near the tire when filling, there is usually no time to react if there's an accidental explosion due to over-inflation. Determine the correct inflation amount before filling the tire and do not exceed it. The tire pressure molded on the sidewall of the tire is maximum allowable and not the suggested running pressure. If you fill to the maximum pressure when the tire is cold, you're already running at the edge of safety. Tires and the air inside usually heat up during operation, so air pressures may rise to very dangerous levels. If you don't want to get a face full of rubber, don't fool around with incorrect tire pressures!

FIRE READINESS

Fire is a danger present in any farm workshop and there are a lot of combustible materials to accelerate it. Not only is fire a danger to life and limb, but it could mean the loss of the building and all your tools. Have effective firefighting tools, clearly signify where they are, and know how to use them.

"Getting away from it all" by moving to a small farm usually also means getting farther away from emergency services, such as firefighting services. Even where there is rural volunteer or professional firefighting service, factors such as longer travel distances may increase response time. If your access road is not clearly marked or hard to use for large vehicles, this may add to the difficulty of fighting a workshop fire.

- Be sure all containers for flammable and combustible liquids are clearly and correctly marked so that they do not end up near sources of heat or sparks.
- Never store fuel in breakable food or drink containers.
- Watch out for leaks or deterioration in fuel storage and delivery equipment.
- Before cutting, welding, or soldering a fuel tank, completely remove any vapor or liquid. Explosion and fire can result from using a torch on a tank that was thought to be empty.
- Keep a fire extinguisher in every building in a practical and easy-to-access location so that if a fire develops, you don't spend crucial time running a long way to find an extinguisher.

Whenever there are flammable vapors or dust present, there is a risk that a spark could set them off. Explosion-proof switches eliminate one source of sparks and make your shop look more professional.

• Make sure the fire extinguishers you buy match the type of fire that can be expected in your situation. If you use the wrong unit on a fast-moving fire, you can cause the fire to spread even faster. Read the label on the fire extinguisher and get advice from where you buy it.
• Study how to use the extinguisher so you're ready before a fire ever starts.
• Make sure all extinguishers are serviced at intervals not to exceed one year.
• Know your limits and always think safety first. Fire extinguishers cannot do the job of a local fire department. When a fire burns for more than a couple of minutes, the heat starts to build up and intensify. Once that happens you are past a point of single-handed extinguishing. Get out of the building and let firefighters handle it safely.
• Be ready to give the fire dispatcher clear, concise directions to your farm. Don't use any instructions such as "turn left where the Smith's barn used to be" or similar confusing terms. Add GPS coordinates if you can provide them.
• Make sure that the access roads or trails to farm areas are smooth and can support large vehicles. Many farms have only a rutted, bumpy trail to the yard, which makes it difficult to move firefighting equipment in and out. If a truck bringing gravel to your yard has a hard time getting in and turning around, firefighting equipment can have as much or more trouble.
• There should be a reliable, accessible water source, such as a pond, no farther than 100 feet from major buildings. Fires can happen in winter so don't count on seasonal water flows or bodies of water that ice up heavily.
• Have a list of what flammable items are stored where so fire fighters know what to be prepared for and what to protect themselves against. Location of fuels, lubricants, and pesticides are especially important.
• Firefighters may ask you for priorities on which building to save. Have a plan on whether it's your shop, barn, machinery storage, or house. If a shop is already burning fiercely, it may be safer and more realistic to save an adjacent building or vehicles stored in it.

GASES, CHEMICALS, AND DUST

A farm workshop, even though it's out in the country, has the same air quality issues as an industrial shop or factory. For example, welding fumes from extensive work in a poorly ventilated shop can contribute to the development of Parkinson's disease, which affects the brain and nervous system and causes symptoms including uncontrollable tremors (shaking), slowness of movement, stiffness, difficulty with balance, and depression. Paints and solvents can also give off various hazardous and/or flammable gases as they dry so be sure to work with proper protective equipment and plenty of ventilation when using these materials.

Carbon monoxide (CO) is a clear, odorless, and very deadly gas formed during combustion in equipment such as internal combustion engines and space heaters. Especially if you are taking over an older, less maintained farm, watch out for areas where CO may collect, such as in shops with old, poorly maintained heaters. Equip your shop with a CO monitor available at building supply centers and discount stores.

Before bringing machinery into the shop, clean it thoroughly with water and/or the pressure washer. Working on dusty equipment can lead to several nasty medical conditions associated with agricultural dusts and molds. Farmers

Any light over a workbench should be protected with a cage. If you accidentally shatter the bulbs while working, flying shards of glass could result.

Lung (also known as extrinsic allergic alveolitis or hypersensitivity pneumonitis), is an incurable, allergic lung disease caused by the inhalation of spores found in moldy crops, such as hay, straw, corn, silage, grain, tobacco, or in bird-breeding and mushroom-growing operations. Exposure is marked by increasing shortness of breath, occasional fever, loss of weight, and general lack of energy. The victim may eventually suffer permanent lung damage or death.

Organic Toxic Dust Syndrome (also known as pulmonary mycotoxicosis), Silo Unloader's Syndrome (SUS), or Atypical Farmers Lung, a flu-like illness due to the inhalation of grain dust with symptoms including fever, chest tightness, cough, and muscle aching, are other medical conditions that plague farmers.

Many different chemicals may be encountered on a farm, even if no pesticides are purchased or used. Proper use, storage, and disposal of chemicals protect yourself and others from accidental exposure. Proper storage and disposal are particularly important for protecting inquisitive children exploring the farm workshop.

Whether you're using workshop chemicals or moving them to a proper disposal site, it's important to follow the manufacturer's recommended handling and application procedures and to wear correct personal protective gear such as respirators, eye protection, impermeable aprons, gloves, and so on.

Farm and rural area safety information is a key mission of the continental network of agricultural extension services. Your local branch can be found in the telephone directory or at these sites:

South Dakota's State Departments of Agriculture:
www.state.sd.us/doa/department
Canadian Provincial and Territorial Departments
of Agriculture: www.agr.gc.ca/index_e.php
Chemical/Pesticide topics:
www.cdc.gov/nasd/menu/topic/chem_general.html

APPENDIX

Commonly Used Tools and Supplies

Dust masks
Eye protectors
Eye wash kit
Fire extinguisher(s)
First-aid kit
Hearing protectors
Heavy-duty hand cleaner
Heavy-duty work gloves
Large bucket of sand for smothering fires
Shop towels

Air tools
Angle grinder
Blow gun
Chisel and bit set
Compressor
Die grinder
High-volume blow gun
Impact wrenches, various sizes, with impact sockets
Nibbler
Panel crimper
Ratchet
Sandblaster
Saw
Shear

Electric tools
Angle grinder (corded and cordless) and wheels
Band saw for metal
Battery charger
Battery load tester
Cutoff saw and spare wheels
Grinder and wheels
Hand drills (corded and cordless), bits, and step drills
Heat gun
Parts washers
Portable lights
Press drill (benchtop and floor type)

Hand tools
Chisels/Punches
Automatic center punch
Bolt cutters
Cold chisel set
Hollow punch set
Measuring tapes, rulers, pencils, felt-tip markers
Non-sparking brass punches
Pry bars
Punch set, including straight, taper, and center punch types
Scriber, carbide tip for marking out metal

Cleaning Supplies
Shop vac
Cleaning brushes

Files
Auger bit file
Emery cloth roll
Gasket scraper
Hacksaw and blades
Machinist's file set
Needle file set
Tap and die set
Thread restoring files

Gauges/Meters
Analog multimeter
Battery hydrometer
Calipers
Coolant hydrometer
Digital multimeter
Feeler gauges
Micrometers
Sheet metal gauge
Spark plug gap tools
Strobe timing light
Tach/dwell meter
Thread gauge

Jacks/Ramps
Chain hoist
Jacks, mechanical and hydraulic types
Jack stands
Vehicle ramps
Wooden blocks

Hammers
Anvil
Ball-peen hammers, various sizes
Body working hammers and dollies
Club hammer, large
Cross-peen hammer, large
Dead-blow hammer
Double-face soft mallet
Pullers, internal and external, various sizes
Slide hammer
Vise

Pliers/Snips
6-inch slip-joint pliers
Aviation snips
Lineman's pliers
Lock ring pliers, inner and outer
Locking pliers, including needle-nose, wide-jaw, and welding clamp types
Needle-nose pliers
Tin snips
Wire connector crimper
Wire cutters, side-cut and end-cut types

Screwdrivers/Knives
Screwdriver set, including blade-, cross-, and square recess–types
Tamperproof screwdriver set
Torx screwdriver set
Utility knife

Wrenches
Combination wrench set
Double-ended open-end wrench set
Box-end wrench set
Stubby wrench set (for tight quarters)
Mini wrench set (for small-size fasteners)
Flare nut wrench set
Crowfoot wrench set
Magnetic pickup tool, to retrieve dropped fasteners
Adjustable wrenches: 6-inch, 8-inch, and 15-inch sizes
Oil filter wrenches
3/8-inch drive torque wrench (10–80 ft-lbs)
1/2-inch drive socket set
1/2-inch drive torque wrench (100–150 ft-lbs)
1/4-inch drive socket set
1/4-drive torque wrench (in-lbs)
3/4-inch, 1-inch drive socket sets (for heavy machinery)
3/8-inch drive socket set

Welding equipment
Arc (stick) welder and electrodes
Arc welding helmet/mask, spare lenses
Gas welding goggles
Leather welding apron/coat and gauntlets
Oxyacetylene welder, rods, striker, and tip cleaner
Plasma cutter
Soldering gun, solder, and flux
Ventilation fans
Water for quenching
Wire feed welder (MIG/TIG), wire, and gases

INDEX